Comparative Genomics and Proteomics in Drug Discovery

T0203735

EXPERIMENTAL BIOLOGY REVIEWS

Comparative Genomics and Proteomics in Drug Discovery

Edited by

JOHN PARRINGTON
Department of Pharmacology, University of Oxford, Oxford, UK

KEVIN COWARD
Department of Pharmacology, University of Oxford, Oxford, UK

CRC Press
Taylor & Francis Group
Boca Raton London New York

CRC Press is an imprint of the
Taylor & Francis Group, an **informa** business

CRC Press
Taylor & Francis Group
6000 Broken Sound Parkway NW, Suite 300
Boca Raton, FL 33487-2742

First issued in paperback 2019

ISBN-13: 978-0-415-39653-0 (hbk)
ISBN-13: 978-0-367-38973-4 (pbk)

Library of Congress Cataloging-in-Publication Data

Comparative genomics and proteomics in drug discovery / edited by John Parrington and Kevin Coward.
    p. ; cm. -- (Experimental biology reviews) (SEB symposium series ; v. 58)
    Includes bibliographical references.
    ISBN 0-415-39653-0 (alk. paper)
  1. Pharmacogenomics. 2. Proteomics. 3. Drug development. 4. Pharmacognosy. I. Parrington, John. II. Coward, Kevin, 1969- III. Series. IV. Series: Symposia of the Society for Experimental Biology ; no. 58.
    [DNLM: 1. Pharmacogenetics. 2. Drug Design. 3. Genomics--methods. W1 SY432K no.58 2006 / QV 38 C737 2006]
RM301.3.G45C66 2006
615′ .7--dc22   2006029232

Editor: Elizabeth Owen
Editorial Assistant: Kirsty Lyons
Production Editor: Simon Hill
Typeset by: Keyword Group, UK
Printed by: Cromwell Press Ltd

Contents

Contributors

Sergio Callejas and Sara Melville, University of Cambridge, UK

CS Peacock, Wellcome Trust Sanger Institute, UK

Maria M. Mota, Miguel Prudêncio and Cristina Rodrigues, Universidade de Lisboa, Portugal

David B. Sattelle, Andrew K. Jones, Laurence A. Brown, Steven D. Buckingham, Christopher J. Mee and Luanda Pym, University of Oxford, UK

Jeff Clare, GlaxoSmithKline, UK

John Bilello, GlaxoSmithKline, USA

David Wishart, University of Alberta, Canada

Preface

This book arose from a one-day symposium arranged as part of the Annual Meeting of the Society for Experimental Biology (UK) in Barcelona in 2005 and takes up the important question of how emerging genomic and proteomic technologies are making significant contributions to global drug discovery programmes, and in particular the key role that comparative genomic and proteomic strategies play.

Rapid progress in our understanding of cellular and molecular biology has led to the field of biomedical science undergoing a major revolution over the last 30 years or so. Of particular note is the dramatic development of genomic and proteomic technologies and their associated application in biology and medicine.

Genomics is the study of an organism's genome and involves isolation, identification and mapping of genes along with associated functional dynamics and interaction. The Human Genome Project has revealed that the human genome is composed of 20,000 to 25,000 genes. However, since each gene can be translated into a variety of different proteins via a variety of cellular and molecular mechanisms, it is estimated that the human proteome consists of approximately 1,000,000 differently modified proteins. This in turn means that there are significantly more potential biomarkers or drug targets to be discovered using proteomic approaches rather than genomic approaches. Expression of these proteins is known to vary in response to environmental change and is related to genetic history. Unlike variation in gene expression, any change in protein expression usually results in alteration of function. Genomics and proteomics are thus considered as being highly complementary approaches to the molecular study of disease.

Written by widely respected authorities from both academic and pharmaceutical backgrounds, this book is composed of seven concise chapters. In Chapter 1, Sara Melville (Cambridge University, UK), introduces the use of comparative genomics in drug discovery in one of the three main human pathogens associated with kinetoplastids, the trypanosomatids, parasitic protozoa responsible for a wide range of diseases including African sleeping sickness, Chagas disease and leishmaniasis. In Chapter 2, Chris Peacock (Wellcome Trust Sanger Institute, UK) describes how comparative genomics is being used to assist in developing treatments against the kinetoplastids, a remarkable group of organisms that include major pathogens responsible for thousands of deaths each year and serious illness to millions. Maria Mota (Universidade de lisboa, Portugal) then considers the relevance of host genes in malaria (Chapter 3), a devastating disease that affects extensive areas of Africa, Asia and South/Central America causing up to 2.7 million deaths per year. In Chapter 4, David Sattelle (University of Oxford, UK) describes how comparative genomics has contributed to the study of nicotinic acetylcholine receptors as drug/chemical targets for a number of conditions including genetic and autoimmune disorders. Chapter 5 (Jeff Clare, GlaxoSmithKline, UK) discusses how genomic and proteomic strategies are being used in gene family-based drug discovery projects aimed at isolating novel sodium channel inhibitors. A sub-population of voltage-gated sodium channels are expressed primarily in nerves involved in pain signalling; these are hence of major interest to the pharmaceutical industry as potential targets for improved analgesics. In Chapter 6,

John Bilello (GlaxoSmithKline, USA) discusses how information resulting from genomic and proteomic studies can be translated into disease understanding and effective management of therapy. Finally, in Chapter 7, David Wishart (University of Alberta, Canada) discusses how advances in genomic and information technology have led to the possibility of *in silico* drug target discovery.

The purpose of this book is to provide an introduction to the concepts behind the dynamic and powerful fields of comparative genomics and proteomics and their specific application in drug discovery. To some extent, the book assumes knowledge of basic molecular biology and is targeted at students, researchers and academics in related areas of biomedicine and pharmaceutics and to a more general readership interested in specific applications of genomic and proteomic technologies.

We would like to thank all of the contributing authors and the Society for Experimental Biology (SEB, UK) for helping us to put this book together. We would also like to extend our thanks to Elizabeth Owen and Kirsty Lyons at Taylor & Francis for their extensive assistance in the final compilation of this book.

Oxford, April 2006.   John Parrington & Kevin Coward

Comparative genomics and drug discovery in trypanosomatids

Sergio Callejas and Sara Melville

Trypanosomatids are parasitic protozoa responsible for a range of diseases, including African sleeping sickness, Chagas disease and leishmaniasis. They infect millions of humans worldwide and are responsible for thousands of deaths every year. There are no vaccines and the available drugs, most of them developed many decades ago, have multiple often toxic side-effects. The genome sequences of three of these trypanosomatids (*Trypanosoma brucei*, *T. cruzi* and *Leishmania major*) have been published recently. This allows researchers to compare these genomes and search for common, as well as species-specific, features in order to identify putative drug targets.

In this chapter we describe the main characteristics of trypanosomatids at the genomic level as well as the most relevant findings in terms of comparative genomics. We review current approaches to drug discovery in trypanosomatids that use information derived from the genome project and evaluate the future potential of the genome information available.

1 Introduction

1.1 *General characteristics of trypanosomatids*

1.1.1 Taxonomy

The family Trypanosomatidae belongs to the protist order Kinetoplastida and includes a wide range of flagellated protozoa exhibiting different shapes, life cycles, geographical distributions and cellular, molecular and genetic characteristics. Kinetoplastids are among the most ancient eukaryotes and, according to rRNA studies, they have rRNA lineage extending further back than those of animals, plants or fungi (Fernandes *et al.*, 1993; Beverley, 1996).

As members of this order, all the trypanosomatids present a unique and characteristic structure called the kinetoplast, a network of mitochondrial DNA located within the typical single long mitochondrion of these organisms, which is associated with the basal body of the flagellum. It contains two different kinds of concatenated circular

Comparative Genomics and Proteomics in Drug Discovery, edited by John Parrington and Kevin Coward.
© 2007 Taylor and Francis Group.

DNA molecules called maxicircles and minicircles. Further description of the mitochondrial DNA is presented in Section 2.1.

Two of the genera included in the Trypanosomatidae family are *Trypanosoma* and *Leishmania*. These two genera appeared long ago and some studies support the idea that their common ancestor disappeared between 400 and 600 million years ago (Stevens and Gibson, 1999; Overath *et al.*, 2001; Stevens *et al.*, 2001). Apart from the presence of the kinetoplast, members of these genera present other unique characteristics such as eukaryotic polycistronic transcription, presence of unique organelles (e.g. glycosomes), distinctive metabolic pathways as well as other significant characteristics, not unique to trypanosomatids, such as antigenic variation of surface glycoproteins and expansion/contraction of subtelomeric regions.

Trypanosomatids have in common that they are all parasitic flagellated protozoa. Among them are the causative agents for several diseases; some of the most representative and well studied are *Trypanosoma brucei*, which causes African trypanosomiasis or sleeping sickness, *Trypanosoma cruzi*, which produces American human trypanosomiasis or Chagas disease in South America and *Leishmania major*, responsible for cutaneous leishmaniasis.

1.1.2 Ecology and geographical distribution
The three species mentioned above require a vector to infect new hosts (mammals, including humans), therefore their geographical distribution and ecology is driven by distribution and ecology of their vectors and hosts. In the case of *T. brucei* the vector is the tsetse fly (genus *Glossina*) which is found in Sub-Saharan Africa. Tsetse flies live in habitats that provide shade for developing puparia and resting sites for the adults. Temperature and therefore climate and altitude are determining factors for the development of tsetse fly populations (Hargrove, 2004; Rogers and Robinson, 2004). *Figure 1A* shows the distribution of human sleeping sickness in Africa.

Despite differences in its life cycle (see below), *T. cruzi* shares some common features with *T. brucei*. For instance, it is transmitted from host to host by insects, not by a fly but by members of the order Hemiptera, known as kissing or 'assassin' bugs. Geographical distribution of *T. cruzi* covers areas from southern South America to southern North America (*Figure 1B*).

Geographical distribution of *Leishmania* is wider compared to *Trypanosoma* species as it can be found in eastern and northern Africa, Central and South America, the Middle East, parts of Asia and even in southern Europe. Depending on the type of vector its habitat ranges from rainforest to arid regions. In general, *Leishmania* is transmitted to humans through the bite of sand flies. *Figure 1C* shows world distribution of *L. major*.

According to The World Health Organization members of the Trypanosomatidae family causing human diseases infect between 15 and 20 million people worldwide (most of whom do not have access to treatment) and are responsible for the death of hundreds of thousands of people each year (WHO, 2002). Most of these infections and deaths take place in poor countries where these diseases have a very important social and economic impact. A good example of this is the case of the genus *Trypanosoma*. Apart from the causative agent responsible for human African trypanosomiasis, this genus includes the species *T. vivax* and *T. congolense*, which are unable to infect humans but are responsible for a similar sickness in mammals (mainly in cattle and other livestock) known as 'nagana'. As a result of this, the following

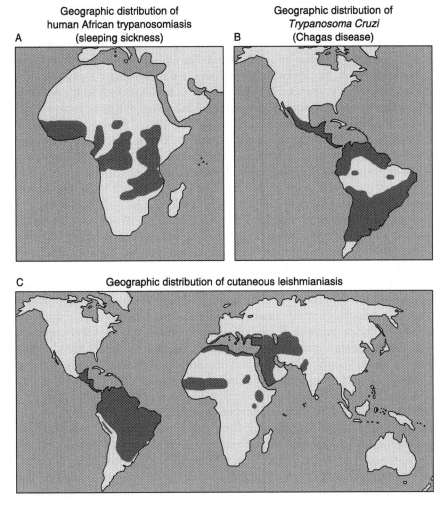

Figure 1. *World distribution of sleeping sickness (A), Chagas disease (B) and cutaneous leishmaniasis (C) caused by* Trypanosoma brucei, Trypanosoma cruzi *and* Leishmania major *respectively. Maps from Atlas of Medical Parasitology, Carlo Denegri Foundation (www.cdfound.to.it).*

observation has been made: on the approximately 7–10 million Km^2 of land infected by the tsetse fly in Africa only 20 million cattle are raised, however, that land could support under other circumstances up to 140 million cattle and increase meat production by 1.5 millions tons (Molyneux and Ashford, 1983).

1.2 *Life cycles*

Life cycles of trypanosomatids are generally very complex and involve more than one host (the vector and a mammal). Here we will describe briefly the life cycle of *T. brucei*, *T. cruzi* and *L. major*, emphasizing the most relevant events that take place during these cycles.

1.2.1 *Trypanosoma brucei*

There are four main forms or stages in the life cycle of *T. brucei*: metacyclic form, bloodstream form, procyclic form and epimastigote. When an infected tsetse fly feeds, the parasite is injected into the mammalian host as a non-dividing metacyclic trypomastigote form. At this stage the parasites are expressing a repertoire of metacyclic variant surface glycoproteins (VSGs) that coat the surface of the parasite, and contain a central vesicular nucleus, a posterior kinetoplast and basal body, with a flagellum running anteriorly along an undulating membrane. As soon as the parasite is in the bloodstream it undergoes a series of morphological and molecular changes to produce the slender bloodstream form of the parasite. This form replicates asexually by binary fission and the flagellum extends beyond the anterior end.

An important molecular event that takes place at this stage is the expression of a new type of VSG. The parasite is able to vary the VSG coat as it replicates, to evade the host immune system (Borst, 2002; Pays *et al.*, 2004). At any one time each cell expresses only one specific VSG (about 10^7 copies per cell forming a dense coat throughout the cell surface) but after a period of time (around 100 divisions) the parasite starts to express a different VSG. Therefore, the host immune system needs to produce new antibodies against the new VSG. This process is repeated to produce a series of parasitaemic waves in the host. This strategy allows the parasite to remain extracellular during its life cycle in the host mammal bloodstream. Eventually, in the last stages of the sickness the parasites also enter the central nervous system (see Section 1.3).

In the bloodstream a transition takes place in a subset of slender form parasites to produce the stumpy form. This form is non-proliferating and appears at the peaks of parasitaemic waves, is preadapted to survival in the tsetse fly but is still covered by VSGs. A tsetse fly becomes infected by taking up stumpy forms when feeding on an infected mammal. In the fly the parasite migrates to the posterior part of the midgut and differentiates into a proliferating form called the procyclic, which divides by binary fission. At this stage the kinetoplast has moved to a region between the nucleus and the posterior end of the cell and the cellular surface has lost its VSG coat and is now covered by a coat of the insect stage-specific protein procyclin. Then the parasite leaves the midgut and migrates to the salivary glands where it differentiates into epimastigotes that continue dividing by binary fission. Epimastigotes have a prenuclear kinetoplast and basal body. Finally, epimastigotes differentiate into metacyclics, ready to be transferred to a new host during the next fly bite. The cycle in the fly takes around three weeks and genetic exchange can occur but is not obligatory.

1.2.2 *Trypanosoma cruzi*

The mammalian host becomes infected during or after the blood meal of an infected triatomine insect vector. These insects release metacyclic trypomastigotes in the faeces, which enter the host through the bite wound or mucosal membranes. Unlike *T. brucei*, once inside the host *T. cruzi* invades cells in different tissues and differentiates into amastigotes, losing the flagellum. Since they are intracellular they are less exposed to the host immune system and do not apply the dramatic phenomenon of antigenic variation. Amastigotes divide by binary fission, differentiate into trypomastigotes and are released in the bloodstream again. In contrast to *T. brucei*, these bloodstream trypomastigotes do not replicate.

The insect vector is infected when it feeds on blood of infected mammals with circulating parasites. The parasites transform into epimastigotes in the vector's

midgut where they divide until they transform into infective metacyclic trypomastigotes in the hindgut.

1.2.3 *Leishmania major*

In all species of *Leishmania*, there are two main stages: the promastigote and the
amastigote. As described for *Trypanosoma* species its life cycle includes a mammalian
host and a vector, in this case a sandfly.

The mammalian host becomes infected with promastigotes during the bite of an
infected sandfly. Once inside the host, the parasites are phagocytosed by macrophages
where they transform into amastigotes and divide rapidly. The very high division rate
shown by these amastigotes is believed to be a strategy to overwhelm the host
immune system. At this stage they have lost the flagellum, as observed in *T. cruzi*.
Eventually, infected macrophages burst and release the amastigotes that will infect
new cells, spreading the infection throughout the host tissue. When an insect vector
bites an infected mammal it swallows infected macrophages containing amastigotes
that, once inside the insect, will differentiate into promastigotes, migrate to the
midgut, divide asexually and eventually transform into the infective promastigote
form. They will be transferred to another host in the next sandfly bite.

1.3 *Pathogenesis*

1.3.1 Human African trypanosomiasis

African sleeping sickness or African trypanosomiasis due to infection with *T. brucei*
occurs in two differentiated stages. The first one presents symptoms associated with the
presence of the parasite in the bloodstream, lymphatic and other body systems but not in
the central nervous system (CNS). The second stage involves damage in the CNS due to
the invasion of the parasite. This second stage may take place as soon as a few weeks after
the infection, or may occur after several years, depending on the *T. brucei* subspecies.

One of the first symptoms is the presence of a chancre at the site of the tsetse bite
that usually disappears within 2–3 weeks. Once the parasite has started to divide in the
bloodstream, blood composition becomes disturbed and attacks of high fever associated with the parasitaemic waves, headache, joint pains and swollen tissues are typical.
The spleen becomes enlarged and there may be signs of heart and intestinal damage,
skeletal muscle atrophy and neurological and endocrine disorders. Once the parasites
reach the CNS additional serious symptoms appear such as mental deterioration, disruption of sleeping patterns, coma and eventually death (Pentreath and Kennedy, 2004).

Symptoms and kinetics of trypanosomiasis may vary depending on the subspecies
that is causing the infection. There are two subspecies of *T. brucei* that can infect
humans, *T.b. rhodesiense* and *T.b. gambiense*. *T.b. rhodesiense* causes an acute disease
that can kill within weeks or months. Populations of this subspecies are found mostly
in eastern and southern Africa. *T.b. gambiense* causes a chronic disease that may take
months or many years to progress to CNS involvement. It is found mostly in western
and central Africa.

Although there are no vaccines, some drugs are available for the treatment of sleeping
sickness; however, their efficiency is not always as high as desired, particularly in the final
stages of the disease, and in most cases side effects are considerable. An additional problem is that most people who need treatment belong to the poorest populations in the
world and cannot always afford it. Some of the registered drugs used for the treatment of

sleeping sickness are: Pentamidine and Suramin for the first stage of the sickness and Melarsoprol and Eflornithine for the second stage (Burri *et al.*, 2004).

1.3.2 American trypanosomiasis

The pathogenesis of American trypanosomiasis or Chagas disease due to infection with *T. cruzi* can be also divided into two phases: acute and chronic. One of the first symptoms is a small sore at the site of the bite. Within days fever appears and lymph nodes swell. This is the acute phase and may cause serious illness and death, especially in young children. However, most patients survive and enter a symptomless and chronic phase for months or years while the parasites invade different organs causing heart, intestinal and oesophageal damage. Around 32% of these patients die as a result of damage suffered during this phase (Lopes and Chapadeiro, 2004).

Despite their toxic side effects, mainly two drugs have been used to treat American trypanosomiasis: Nifurtimox and Benznidazole. Both have been used in the acute and the chronic phase of the sickness with different levels of success (Rassi and Luquetti, 2004). Apart from chemotherapy, prospects of an effective vaccine are very low and one of the most successful approaches to control of this disease is still the control of the vector through treatment of homes with residual insecticides and improvement of building construction to reduce vector habitats.

1.3.3 Leishmaniasis

There are around 20 infective species and subspecies of *Leishmania*, and the range of symptoms and pathogenesis is very wide. The most serious form is visceral leishmaniasis, caused by widespread infection in the lymphatics with *L. donovani*. Typical symptoms include irregular bouts of fever, substantial weight loss, swelling of the spleen and liver, and anaemia. Untreated visceral leishmaniasis has a mortality rate of almost 100%. However, the most common form is cutaneous leishmaniasis due to infection with *L. major*. In this case symptoms are obvious in the skin where ulcers develop, ranging from 1 to 200 in number. These ulcers usually self-heal but leave significant and permanent scars.

Many forms of leishmaniasis are extremely difficult to treat and require a long course of pentavalent antimony drugs. Drug resistance has been reported recently, leading to the use of more toxic drugs. There are no vaccines available and, once again, prevention by using insect repellent or insecticide, control of the vector population based on spraying with residual insecticides, or removal of reservoir hosts are still the most successful approaches to control leishmaniasis.

2 The genomes of trypanosomatids

2.1 *Karyotype and genome organization*

Trypanosomatids have two separate genomes, one in the nucleus and one in the kinetoplast. The kinetoplast is one of the most characteristic cellular structures of trypanosomatids. Indeed, the order Kinetoplastida has taken its name from this structure. However, from the comparative genomics point of view the most interesting and the one that will be reviewed more thoroughly in this chapter is the nuclear genome.

2.1.1 The kinetoplast genome
The kinetoplast is a disc-shaped structure located within the single mitochondrion of trypanosomatids and it contains the mitochondrial DNA known as kDNA. kDNA accounts for 10–20% of total DNA content and is a network of thousands of circular DNA molecules that are topologically interlocked. There are two different types of circular molecules: maxicircles and minicircles.

Maxicircles range from 20 to 40 kilobases (kb) in size depending on the species and encode ribosomal RNA and several proteins, most of them involved in mitochondrial energy transduction. A remarkable feature of these maxicircles is that most of their transcripts undergo RNA editing to form a functional RNA (i.e. addition and deletion of uridine residues).

Minicircles range between 0.5 and 2.9 kb in size and are the most abundant molecules in the kinetoplast: up to 90% of the kinetoplast mass. The sequence analysis of a number of minicircles shows that these molecules do not encode any proteins. Instead they encode guide RNA (55–70 bp), which plays a central role in the editing process mentioned above. According to the available data, it seems clear that production of mature RNAs involves multiple editing events involving many different guide RNAs.

For additional information about trypanosomatid kinetoplasts, some reviews are available (Shapiro and Englund, 1995; Melville *et al.*, 2004; Shlomai, 2004).

2.1.2 The nuclear genome
The organization of the nuclear DNA in trypanosomatids varies substantially between genera and species. In the case of *T. brucei* the haploid nuclear genome is around 35 Mb and it is distributed into three different size classes of chromosomes: minichromosomes, intermediate chromosomes and megabase chromosomes. These different types can be separated using pulsed field gel electrophoresis (PFGE) (Van der Ploeg *et al.*, 1984). Due to lack of chromosomal condensation during mitosis in trypanosomes, PFGE has been the most useful technique to determine the karyotype of these parasites.

T. brucei minichromosomes are linear DNA molecules ranging from 50 to 150 kb in size. There are about 100 of them and they contain *VSG* genes although no expression sites (ES) have been found in minichromosomes. ES are subtelomeric regions that contain *VSG* promoters and therefore are essential for *VSG* expression. It is thought that none of the *VSGs* present in minichromosomes are expressed unless they either undergo an interchromosomal duplication or are part of a telomere exchange. The central core of minichromosomes consists of tandem arrays of a 177 bp repeat that in some cases may account for up to 90% of the whole minichromosome sequence. This core region is a repetitive palindrome, an arrangement also observed in intermediate chromosomes and in small chromosomes in *Leishmania* (Fu *et al.*, 1998; Wickstead *et al.*, 2004). It has been proposed that they serve as repositories for silent *VSGs* for use in antigenic variation. If this is the case, it is surprising that such a large portion of the genome (10–20%) is dedicated just to this purpose (Donelson, 1996).

The intermediate chromosome population consists of several molecules of uncertain ploidy and content ranging from 200 to 900 kb in size. These are the least characterized chromosomes in *T. brucei* and it has been proposed that they may have a similar function to the minichromosomes as they possess *VSGs* or *VSG*-like sequences, but they may also carry expression sites (Berriman *et al.*, 2002).

Finally, there are 11 pairs of megabase chromosomes (Melville *et al.*, 1998) that account for 80% of the nuclear genome and contain housekeeping genes. The megabase chromosomes are numbered from I (the smallest) to XI (the largest) and letters a and b are used to identify homologues. The size of these chromosomes is very variable between *T. brucei* stocks (see Section 2.2) but in the case of the stock selected for the genome sequencing project (TREU927/4) they range from approximately 1.1 to 5.2 Mb (Melville *et al.*, 1998). Hybridization of pulsed field gels with specific probes has also shown that some megabase chromosomes may carry *VSG* expression sites active in bloodstream infection. Current data support the idea that such ES may be found on one homologue but not on the other while some chromosome pairs have ES on both homologues. Distribution of ES is not conserved between different stocks of *T. brucei* suggesting that recombination between homologous sequences on non-homologous chromosomes may occur allowing the transfer of ES between chromosomes. Because ES are located in the subtelomeric regions of the megabase chromosomes and their distribution is so variable, such subtelomeres are considered segmentally aneuploid (Melville *et al.*, 1998, 1999).

Unexpressed copies of *VSG* genes have also been found in the megabase chromosomes. According to Southern blot analysis with *VSG* probes these *VSGs* are haploid, although this may be due to the high level of sequence divergence between individual alleles (Melville *et al.*, 2000). However, analyses of chromosomal polymorphisms carried out in *T. brucei* 427, where the difference in size between the two chromosome I homologues is 1.75 Mb, show that the most likely cause of size variation is the presence of large *VSG* arrays in the biggest homologue, suggesting these *VSGs* are in fact haploid and the arrays are also in segmental aneuploid regions of chromosomes (Callejas *et al.*, in press). Nevertheless, generally speaking and apart from the minichromosomes, intermediate chromosomes and the subtelomeric region of megabase chromosomes, the nuclear genome of trypanosomes is diploid (Melville *et al.*, 2004).

In *T. cruzi* the haploid nuclear genome is around 55 Mb and is distributed in approximately 28 pairs of chromosomes, although the exact number is unknown. According to available data from PFGE, the genomic DNA of *T. cruzi* can be divided into 12 megabase bands (each band may contain more than one chromosome) ranging from 3.5 to 1.0 Mb and 8 intermediate bands ranging from 1.0 to 0.45 Mb (Cano *et al.*, 1995; Santos *et al.*, 1997; Porcile *et al.*, 2003). However, no bands smaller than 0.45 Mb that could be interpreted as minichromosomes were detected.

Data from Southern blot hybridizations of PFGE with different gene probes have shown that they mostly hybridize to a unique band, or two bands, suggesting size polymorphisms between homologous chromosomes (Cano *et al.*, 1995). Hundreds of genetic markers have since been mapped to the chromosomal bands of *T. cruzi* including probes originating from protein-coding sequences, structural RNAs, cDNAs, ESTs (Expressed Sequence Tags) and GSSs (Genomic Survey Sequences) (Santos *et al.*, 1997; Porcile *et al.*, 2003). Nevertheless, the complexity of the karyotype, including trisomy and variation between strains, has so far prevented complete resolution.

The haploid nuclear genome size of *L. major* is 33 Mb and presents 36 pairs of chromosomes ranging from 0.28 to 2.8 Mb (Wincker *et al.*, 1996). *L. major* is a diploid organism but some chromosomes may be aneuploid (Scholler *et al.*, 1986; Bastien *et al.*, 1998). The overall molecular karyotype of *Leishmania* is conserved among the different strains and species (Bastien *et al.*, 1992) although some size polymorphism exists between them. The structure of *L. major* chromosomes consists of central regions similar to those found in *T. brucei* and *T. cruzi*, with a high density of coding sequences (Myler *et al.*, 1999), and small subtelomeric regions that contain mostly non-coding repetitive sequences (Ravel *et al.*, 1995; Fu and Barker, 1998; Myler *et al.*, 1999).

2.2 *Chromosome polymorphism: subtelomeric regions*

Although the extent varies, all trypanosomatids exhibit some degree of chromosome size polymorphism between species, subspecies, strains and even between homologous chromosomes within strains. In *Leishmania* chromosome size variation of up to 25% has been observed (Wincker *et al.*, 1996) and in *T. cruzi* two-fold variation has been reported (Henriksson *et al.*, 1995). But probably the most remarkable examples occur in African trypanosomes where apparently homologous chromosomes (i.e. containing the same coding sequences and, where investigated, in the same order) vary up to four-fold in size (Melville *et al.*, 1998, 2000). Nevertheless, no major DNA translocations have been detected so far in these chromosomes (Melville *et al.*, 1998, 1999, 2000).

The underlying cause(s) of these chromosome size polymorphisms remains unclear but some events have been suggested, such as the variable location of ES and *VSGs* in different stocks, variation in the length of repeated sequences associated with ESs, variation in the number of transposon elements in the genome or variation in the number of copies of genes in tandemly repeated arrays (Callejas *et al.*, in press; Melville *et al.*, 1999).

Whatever the reason for chromosome size polymorphisms in these parasites it seems that most of it is due to variation in subtelomeric regions. This could be a consequence of a common feature observed in trypanosomatids and other parasites, as revealed by the parasite sequencing projects: their chromosome organization, where genes that are conserved and syntenic (i.e. conservation of gene order) in related species are concentrated in the gene-dense chromosome core, while species-specific gene families that diverge more rapidly within populations are located at the ends of chromosomes (El-Sayed *et al.*, 2005a; Hall *et al.*, 2003; Pain *et al.*, 2005). In lower eukaryotes, subtelomeric plasticity is thought to help organisms to adapt to new environmental conditions by allowing rapid evolution of the gene families. This appears to be especially marked in pathogenic organisms (Freitas-Junior *et al.*, 2000; Barry *et al.*, 2003; Fabre *et al.*, 2005).

A recent comparison of the sequenced chromosome I of *T. brucei* strain 927 (1.1 Mb) (Hall *et al.* 2003) with the much larger chromosome I homologue of strain 427 (3.6 Mb), using comparative genomic hybridization, revealed only minor variation in DNA content in the chromosome core. Over 2 Mb of extra DNA was located in the subtelomeres, i.e. segmentally haploid regions (Callejas *et al.*, in press). Amongst the amplified gene families is the Retrotransposon Hot Spot (*RHS*) family, a diverse set of genes that appears to be evolving rapidly and into which retrotransposons preferentially insert at a specific site (Bringaud *et al.*, 2002b). These genes encode nuclear proteins of unknown function: since most subtelomeric gene families in these parasites appear to encode surface proteins, this remains an intriguing observation. However, the greatest amplification was seen in the *VSG* array, which is massively expanded in the larger chromosome (Callejas *et al.*, in press). This suggests that the number of *VSGs* in the sequenced strain, which has smaller chromosomes than most strains, may be considerably underestimated.

Polymorphism in subtelomeres is observed in *Leishmania* also, although it is less extensive and data published to date suggest that it involves primarily repetitive sequence such as that observed on chromosome I (Sunkin *et al.*, 2000). Variation in *T. cruzi* subtelomeres may be due to a combination of such factors, as they contain large gene families encoding surface proteins, repetitive DNA and clusters of retro-transposons (Henriksson *et al.*, 1996; El-Sayed *et al.*, 2005b).

2.3 *Sequenced genomes*

The genome sequences of the trypanosomatids *T. brucei*, *T. cruzi* and *L. major* have been published simultaneously (Berriman *et al.*, 2005; El-Sayed *et al.*, 2005b; Ivens *et al.*, 2005). This represents a major breakthrough in the study of these parasites, making available massive amounts of information and allowing researchers to compare these three genomes with each other and with other eukaryotic genomes in order to find species-specific genes and metabolic pathways, as well as common features. The availability of these three genome sequences will help us to understand the genetic and evolutionary bases of these parasites.

As mentioned above, one of the common features to these genomes is their chromosome structure: on the one hand, they all show a highly gene-dense central core where the housekeeping genes are located and synteny is conserved between species and, on the other hand, they all have polymorphic subtelomeric regions where repetitive elements and species-specific gene families are usually found.

Despite those similarities, each genome has its specific characteristics. In the case of *T. brucei* one of the most notable features is that around 20% of the genome encodes subtelomeric genes, most of them *T. brucei*-specific, including the *VSGs*. Remarkably, only 7% of analysed *VSGs* are fully functional, the rest of them are defective in different ways. Some are atypical with inconsistent folding or posttranscriptional modification, many are pseudogenes with frameshifts or in-frame stop codons and some are just gene fragments (Berriman *et al.*, 2005). It is unclear why *T. brucei* stores such a large number of defective *VSGs*. It is possible that these represent *VSG* graveyards, i.e. *VSG* copies left behind after recombination events during antigen-switching, and undergoing sequence decay. If that is the case, then most trypanosomes are replicating very large amounts of useless ('junk') DNA during each cell cycle. An alternative explanation is that pseudogenes and gene fragments are in fact used to create novel *VSG* genes by recombination before or during transposition to an ES, as occasionally reported (Roth *et al.*, 1989). Then the potential size of the *VSG* repertoire in the subtelomeric regions suggests an enormous capacity for new antigenic types in African trypanosome populations.

The genome project has provided important data that may help to understand chromosome evolution in these parasites. *T. brucei* has 11 large diploid chromosomes while *T. cruzi* and *L. major* have around 28 and 36 pairs, respectively. In the last two cases these chromosomes are smaller compared with *T. brucei*. It seems that the 11 chromosomes of *T. brucei* may be the result of fusions of the 36 chromosomes from *L. major* (or a common ancestor). This idea is supported by the fact that 20 of the 36 chromosomes of *L. major* are highly syntenic within a larger *T. brucei* chromosome (El-Sayed *et al.*, 2005a). Apparent chromosomal fusions found within *T. brucei*, where ends of two chromosomes from *L. major* or *T. cruzi* seem to have joined to form a single chromosome in *T. brucei*, also support this hypothesis. The fusion region in *T. brucei* (internal to the chromosome) contains typical telomeric sequences such as *RHS*, DIRE and *ingi*, suggesting a telomeric origin for that region (El-Sayed *et al*., 2005a). It is thought that the current chromosomal structure of *T. brucei* derives from an ancestor with a more fragmented genomic organization similar to that found in *L. major* and *T. cruzi*. *Figure 2* summarizes the typical structure of *T. brucei* chromosomes, as well as syntenic blocks found in *L. major*.

In terms of genome evolution it is interesting to note that no relation between the parasite lifestyle (intra- or extracellular) and the level of genome compaction in these species is observed. It has been suggested that in *Encephalitozoon cuniculi* intracellular lifestyle is associated with a compaction of the genome (Zhang, 2000; Katinka *et al*., 2001). However, gene density in *L. major* (intracellular) is lower than in *T. brucei* (extracellular) or *T. cruzi* (intracellular).

3 Comparative genomics of trypanosomatids

3.1 *Synteny in trypanosomatids*

The term synteny describes the conservation of gene order between different organisms. The study of this conservation is a helpful tool to determine evolutionary relationships between organisms as well as to analyse what kinds of mutational forces have

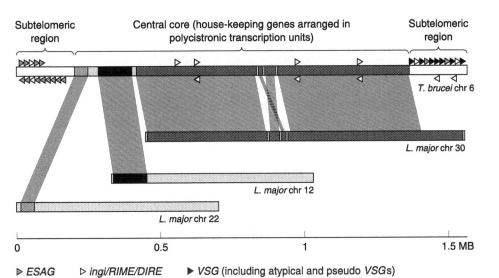

Figure 2. T. brucei *chromosome VI structure showing the typical chromosome organization in trypanosomatids: a central core containing house-keeping genes and subtelomeric regions with variable gene families, most of them involved in antigenic variation. Shaded blocks represent syntenic regions found in* L. major *chromosomes. Triangles do not represent single genes and positions are approximate. Based on data taken from Berriman* et al. *(2005) and El-Sayed* et al. *(2005a).*

driven chromosomal architecture evolution. It is proposed that related genomes where synteny is poorly conserved must have been exposed to stronger mutational forces than those where synteny is highly conserved, or that genomes where synteny is highly conserved have been exposed to strong selection pressure to preserve gene order. Synteny has also proven to be extremely useful in the assignment of orthology of genes across different organisms, since similar genes found in similar positions are usually orthologous.

One of the conclusions derived from the *T. brucei*, *T. cruzi* and *L. major* genome sequencing projects is that, although *Trypanosoma* and *Leishmania* last shared a common ancestor between 400 and 600 million years ago (Stevens and Gibson, 1999; Overath *et al.*, 2001; Stevens *et al.*, 2001), gene order in trypanosomatids is highly conserved. Extensive analyses on different DNA fragments from these three trypanosomatids have been carried out confirming this gene order conservation (Bringaud *et al.*, 1998; Ghedin *et al.*, 2004; El-Sayed *et al.*, 2005a). For instance, more than 80% of the 79 coding sequences present in chromosome I in *L. major* have been found on a 272.5 kb region of *T. brucei* chromosome IX. Sometimes inversions of portions of chromosomes are found between species; however, although inverted, gene order is still conserved. Moreover, of all the genes in *T. brucei* and *L. major*, 68% and 75%, respectively, remain in the same genomic context. And 94% of all the genes that form the proteome core in the three species fall within regions of conserved synteny.

There is still a very high number of open reading frames (ORFs) in these three trypanosomatid genomes that do not show any significant identity with any gene of known function and that have been classified as 'genes' using bioinformatic tools only. These genes are divided into two categories: hypothetical genes, if they are found only in one species, and conserved hypothetical genes, if they are found in multiple species. Hypothetical genes are less likely to be true genes than conserved hypothetical genes. If conserved hypothetical genes fall in regions of synteny, this is also valuable evidence that they are true genes. Conserved hypothetical genes present only in the three species of trypanosomatid may be good candidates for trypanosomatid-specific genes that could be considered potential targets for anti-trypanosomatid drugs. Hypothetical genes present in just one species of trypanosomatids may represent species-specific genes and therefore they may be among the most interesting targets for species-specific drugs.

3.2 *Possible hypothesis for explaining synteny*

An amino acid sequence comparison of regions showing synteny has revealed a high level of sequence divergence between *L. major* and *Trypanosoma* species and even between *Trypanosoma* species (Ghedin *et al.*, 2004). The average identity found at this level between *L. major* and *Trypanosoma* homologues in these regions is 40–45%, while between *T. brucei* and *T. cruzi* the average is approximately 55%. These levels of divergence are within the range expected for lineages that diverged a long time ago, as suggested for *Leishmania* and *Trypanosoma*. However, the extensive synteny observed within the trypanosomatids is well above that expected for these two lineages. As Ghedin *et al.* explain, there has been ample time for the genome to rearrange and the conservation is astonishing when compared to other organisms. For example, in the nematodes *Caenorhabditis elegans* and *Caenorhabditis briggsae* (estimated to have diverged only 25–50 million years ago) it is observed that these two genomes only align approximately 9 kb before an interruption in their gene order (Kent and Zahler, 2000). It seems clear that there must be a strong selective pressure to maintain gene order in trypanosomatids.

Unlike in prokaryotes where gene order is maintained for clusters of genes with a similar function (operons), no evidence has been found in trypanosomes of genes grouped by function (El-Sayed et al., 2005a, 2005b; Ivens et al., 2005) with the exception of the enzymes involved in 'de novo' pyrimidine biosynthesis in T. cruzi (Gao et al., 1999). Also, synteny conservation in trypanosomatids due to lack of homologous recombination seems unlikely since it is known that genetic exchange takes place at least in Trypanosoma. It looks more likely that in trypanosomatids the atypical transcriptional mechanism used by these organisms, eukaryotic polycistronic transcription, could favour synteny. In trypanosomatids there is an unusual lack of identifiable promoters for RNA polymerase II, the enzyme that usually transcribes protein-coding genes in eukaryotic organisms. This lack of promoters may be explained by the fact that genes in trypanosomatids are transcribed into large polycistronic precursor RNAs that are cleaved into monocistronic mRNAs (Donelson et al., 1999). It is unclear the reason for the presence of these polycistronic units but it has been suggested that origins of replication may be located between the polycistronic units (McDonagh et al., 2000) to allow replication and transcription to be co-directional. If this is the case, most genome rearrangements would break these units and would produce an alteration in the coordination of replication and transcription. This would impose a negative selection against any of these rearrangements.

3.3 Synteny breaks: retrotransposon-like elements and subtelomeric regions

Although there is a remarkable overall synteny between trypanosomatid genomes, many insertions, deletions or substitutions have been found within syntenic regions (Ghedin et al., 2004; El-Sayed et al., 2005a). However, these breakpoints of synteny are not distributed randomly throughout the genome; instead, they seem to be situated at specific locations. For instance, all the rearrangements found by Ghedin et al. are associated with strand-switch regions or chromosome ends. Moreover, certain DNA sequences seem to be associated with them, for example non-long terminal repeat (non-LTR) retrotransposon elements. In fact, all the retroelements found in the segments analysed by Ghedin et al. were located in strand-switch regions. The non-LTR retrotransposons may play an important role in chromosome rearrangement, at least in T. brucei and T. cruzi where some of them (ingi and L1Tc respectively) are the most abundant elements described in these genomes. Comparison of the DNA sequences of ingi and L1Tc elements shows that elements within each group are very conserved (Ghedin et al., 2004), supporting the idea that these elements may still be active in the genome or may have been active until very recently. Recombination between these mobile elements may rearrange the genome, perhaps generating the observed strand-switch regions and the chromosomal rearrangements. However, no transposable elements have been detected so far in Leishmania species (Hasan et al., 1984; Kimmel et al., 1987; Aksoy et al., 1990; Villanueva et al., 1991; Martin et al., 1995; Bringaud et al., 2002; Ivens et al., 2005).

The most obvious synteny breaks in trypanosomatids are found in chromosome-internal regions containing tandem arrays of protein-coding and RNA genes (as in

L. major) and in the transition to subtelomeric sequences (as in *T. brucei* and *T. cruzi*) (El-Sayed *et al.*, 2005a). The non-syntenic subtelomeric regions of *T. brucei* may be as large as several hundred kilobases long and, as previously described, consist of arrays of species-specific *VSG* genes with ES-associated genes (*ESAGs*), retroelements and *RHS* genes. In *T. cruzi* subtelomeres are particularly rich in repetitive sequences; however, interesting genes are also found in this region, such as interspersed arrays of the transialidase (surface protein) superfamily, DFG-1 and pseudo *RHS* genes. In addition, it is interesting to note that in *T. cruzi* non-syntenic 'islands' containing genes encoding surface proteins and associated genes seem to be located between chromosome internal synteny blocks, outside the subtelomeric regions (El-Sayed *et al.*, 2005b). In *L. major* subtelomeric organization is slightly different; subtelomeres are smaller compared to *T. brucei* and *T. cruzi* (less than 20 kb) with just a few repetitive sequences. Yet, the most distal genes are often nonsyntenic, although usually not species-specific (Ivens *et al.*, 2005).

3.4 Species-specific genes and domains

Identification and study of species-specific genes in trypanosomatids is one of the most important outcomes of the sequencing project. From the drug discovery point of view the use of species-specific genes as targets for chemotherapy is a very promising approach.

The genomes of *T. brucei*, *T. cruzi* and *L. major* encode around 8100, 12,000 and 8300 protein-coding genes, respectively. Analyses of the predicted protein sequences corresponding to all these genes in the three species have allowed the grouping of these genes into clusters of orthologous genes (El-Sayed *et al.*, 2005a). As result of these analyses 6158 clusters were allocated as common to the three species; these represent the core proteome of these three parasites. Some 1014 clusters were common to only two of the parasites; however, the distribution of these clusters is different depending on which parasites are compared. For instance, more clusters are shared in the two intracellular parasites *L. major* and *T. cruzi* than observed in the two trypanosome species. *T. brucei* and *L. major* share the fewest clusters. Comparison of the amino acid sequence of a large sample of the 6158 clusters shows that the average identity between *T. brucei* and *T. cruzi* is 57% but only 44% between *L. major* and the other two trypanosomes.

There is also an important part of each proteome that corresponds to species-specific clusters. *T. cruzi* is the trypanosomatid that seems to have the highest proportion of these genes (32%) followed by *T. brucei* (26%) and then far behind by *L. major* (12%). Most of these species-specific proteins seem to be members of surface antigen families (involved in the evasion of the host immune system). Since these parasites use different strategies for immune evasion, it is possible that the different numbers of species-specific genes may be related to each parasite-specific strategy for survival in their hosts.

Another way to compare genomes is to look at protein domains that carry out specific functions and identify those that are species-specific. A comparison of the three sequenced trypanosomatids has revealed that of 1617 identified protein domains in the three trypanosomatids, fewer than 5% are species-specific (El-Sayed *et al.*, 2005a). As expected most of these species-specific domains are involved in metabolic and physiological functions unique for each parasite and strongly related to their lifestyles.

One example is the macrophage migration inhibitory factor domain that may inhibit the macrophage activation and therefore the destruction of the parasite. This domain is unique to *L. major*, as well as another protein involved in vacuolar transport, probably diverting proteases within the host phagolysosome. The reason why this domain is not found in the other two trypanosomatids may be because *T. brucei* is extracellular and *T. cruzi* has a very short stay in the phagolysosome before escaping to the cytoplasm, and therefore it is not crucial to avoid the protease activity in that organelle.

Examples of *T. brucei*-specific domains are those identified in the *ESAG1* and *ESAG6-7* genes. These genes are not related to the immune system evasion but to the requirement for iron. *T. brucei* needs iron for its propagation and this is taken from the host transferrin by the parasite through a heterodimeric glycosylphosphatidylinositol-anchored receptor encoded by *ESAG6* and 7 genes (Steverding *et al.*, 1994). Other *T. brucei*-specific domains are the AOX domain that may act as an alternative terminal oxidase in mitochondria and the LigB domain involved in aromatic compound metabolism.

T. cruzi needs to enter host cells in order to spread infection. To do this the parasite secretes a peptide that interacts with a specific host-cell receptor and triggers migration of lysosomes to the host-cell membrane, allowing the parasite to enter the cell. The peptide *T. cruzi* secretes is processed by a serine peptidase so it is not surprising that one of the *T. cruzi*-specific domains found is a serine carboxipeptidase (Burleigh and Andrews, 1998; Caler *et al.*, 1998; El-Sayed *et al.*, 2005a). Some other *T. cruzi*-specific domains have been found such as hormone-type domains, although their significance is not clear yet.

4 Drug discovery in trypanosomatids

4.1 *Introduction*

Despite the differences described between the *T. brucei*, *T. cruzi* and *L. major* life-cycles and the species-specific genes found in particular regions of some chromosomes, there is a remarkable overall similarity between these three genomes. Although there is sequence divergence, for these lineages diverged a long time ago, most of the genes are present in all three species and synteny is extremely highly conserved. However, despite this overall similarity these species cause very different diseases (*Figure 3*). In the late stages of sleeping sickness the extracellular parasites located in the bloodstream cross the blood–brain barrier and penetrate the CNS. Therefore, an effective late-stage drug must be capable of crossing the blood–brain barrier in sufficient concentration. Usually this requires higher concentrations of drug in the bloodstream, which may be toxic for the patient. The intracellular parasite *T. cruzi* may infect almost any cell in the body, and therefore a drug able to reach and enter different cell types at a therapeutic concentration is needed. In the case of *Leishmania*, members of the different species complexes exhibit different tissue tropism and different types of tissue damage, and therefore specific drugs may be necessary to treat the various disease manifestations.

Such variation in pathogenesis complicates the development of effective drugs, and reduces the likelihood that identifying targets that look promising in all three parasites will lead to a drug that is useful against more than one disease or stage.

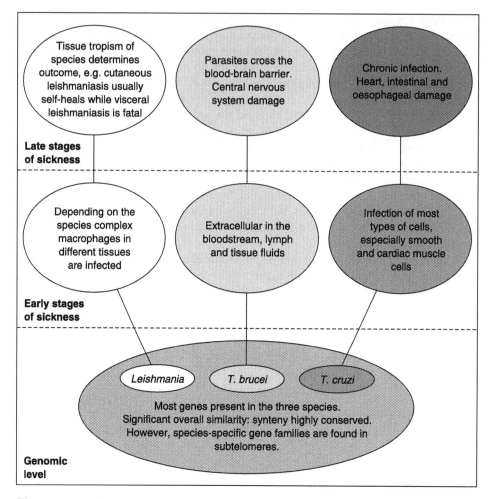

Figure 3. Although similarity among trypanosomatids is relatively high at genomic level, they cause very different diseases, which each pose specific challenges to drug development. The significant variation in pathogenetic mechanisms may preclude development of cross-species drugs.

As discussed previously in Section 1.3 there are no ideal drugs for the treatment of any of the diseases caused by trypanosomatids. Existing drugs produce significant side effects, efficiency may be low and drug resistance has been detected. Vaccines are not available and their development presents considerable difficulties. Few new drugs against sleeping sickness have been developed in the last century. Melarsoprol, an arsenic derivate, kills one in twenty people, although it has been used since 1949. The newer drug eflornithine is less toxic but needs to be injected four times a day, and is not effective against *rhodesiense*. For treatment of Chagas disease there are even fewer drugs available and the most successful approach to combat this disease is still control of the vector. Antimonial drugs for treatment of leishmaniasis are very toxic.

Development of drugs for the treatment of these diseases faces several problems, not only technical but also financial: many drug companies have reduced interest in developing drugs for tropical diseases due to the very little chance of ensuring a return on their financial investment, although there are some recent innovative developments (Moran *et al.*, 2005; Morel *et al.*, 2005; The Wellcome Trust, 2005).

The genome sequencing of *T. brucei*, *T. cruzi* and *L. major* has opened a huge range of new possibilities for drug and vaccine development against trypanosomatids. As we have seen previously, the genomes of these three parasites show common features as well as species-specific genes, domains or metabolic pathways. Both common and unique features are potentially useful. Unique features may provide targets for treatment of a specific disease or stage, while even more interesting is the possibility that a target unique to the family Trypanosomatidae may lead to wide-spectrum drugs, which may be more attractive to drug companies. Nevertheless, it is likely that wide-spectrum drugs will be most effective against early stages of infections due to the differences in terms of pathogenesis caused by each parasite (Figure 3).

4.2 *Strategies*

The perfect drug against trypanosomatids is one that will interact with a component of the parasite, blocking a function or a metabolic pathway essential for its survival, but that will not interact with the host organism, resulting in low toxicity for the mammal. The obvious way to find such drugs is to identify a target that is unique to the parasite (or parasites) but not present in the host.

There are basically two ways of identifying these targets: the classic approach and high-throughput target screening, now more applicable since the development of microarray technology and the availability of multiple genome sequences. The main difference between these two strategies is the number of potential targets that may be screened.

4.2.1 Classic approach

In the classic approach a specific metabolic pathway or enzyme that is thought to be involved in a vital reaction for parasite survival is targeted and inhibitors are selected or developed. Target validation (i.e. verification that the selected target is really essential for the parasite survival) is usually done by inhibiting the expression of the gene encoding the targeted protein. Two of the most commonly used techniques to inhibit gene expression are gene-knockout and RNA interference (RNAi). Gene-knockout consists of silencing the targeted gene by removing it from the genome through transformation and recombination. As each gene is present as diploid alleles, two rounds of knockout are required. On the other hand, RNAi consists of introducing into the cell a plasmid encoding a double-stranded RNA that is homologous to a portion of the targeted gene. The presence of the double-stranded RNA leads to degradation of the endogenous mRNA. The RNAi phenomenon in trypanosomatids was described for the first time in 1998 (Ngo *et al.*, 1998). The main advantages of this technique are that only one round of transformation is required and that it can be used to silence single-locus and multi-locus genes. However, although RNAi has proven to be a very successful technique in *T. brucei*, it seems it does not work in *Leishmania* probably due to lack of some or all components of the RNAi machinery (Robinson and Beverley, 2003).

Target validation is one of the main hurdles during the drug development process and here is where genome sequence information plays an important role. It is important to determine whether similar proteins are encoded in other trypanosomatids and in the human host. Gene sequences provide information about protein domain families, such that inhibitors that bind conserved trypanosomatid-specific domains may be developed. And finally, genome projects provide access to the sequence of the target gene and its environs in the genome to design the RNAi or the knockout experiments.

There are already some good candidate targets for the treatment of some of the diseases caused by trypanosomatids, for example in *T. brucei* where an enzyme involved in the transformation of UDP-Glc to UDP-Gal, named UDP-glucose 4' epimerase, seems to be essential for the parasite. UDP-Gal is necessary for the correct assembly of the parasite coat but the parasite is unable to take it up, so it depends completely on the transformation from glucose (Roper *et al.*, 2002). If it is possible to find an inhibitor for this enzyme that does not interact with the host metabolic pathways it could be used as a drug against *T. brucei*. Another candidate is the glycosylphosphatidylinositol (GPI) biosynthetic pathway. The GPI molecule is necessary to anchor the cell surface molecules; without GPI the parasite loses its coat and is vulnerable. Analogues of a GPI intermediate have proven to disrupt the GPI biosynthetic pathway in *T. brucei* and therefore to be toxic to this parasite (Smith *et al.*, 2004). However, the equivalent human pathway does not metabolize these analogues so they are not toxic for humans.

In some cases common putative targets to different trypanosomatids have been found, such as the enzyme N-myristoyl transferase (NMT). NMT has been characterized in *T. brucei* and *L. major* and it is essential to these two organisms, as gene targeting experiments and RNAi assays have demonstrated (Price *et al.*, 2003). Although this enzyme is not unique to these organisms and is present in humans as well, it seems to be different enough to be affected by any drug designed to inhibit its function in trypanosomatids.

4.2.2 High-throughput target screening

The high-throughput approach to drug discovery in trypanosomatids is still in its early stages since the whole genome sequences have become available only recently. Nevertheless, the opportunities that the genome sequences provide are enormous. We have now the DNA sequence of three complete genomes of trypanosomatids involving thousands of genes (many of them of unknown function) and multiple metabolic pathways. This gives us the opportunity to screen throughout these genomes searching for potential drug targets. To efficiently handle this huge amount of information it is necessary to set up a series of different steps or 'filters' in order to focus on the most interesting genes or regions for drug development. Basically, the most common strategy is as follows.

First, it is necessary to identify all the genes that can be used as potential drug targets. This is done through bioinformatics and data mining. The criteria used for selection may vary depending on the needs, but some features such as genes or metabolic pathways absent or very divergent in the mammalian host (some examples are shown in the metabolic pathways described in Berriman *et al.*, 2005) and genes present and conserved only in trypanosomatids are among the most common. At this stage, genes of unknown function (the majority of annotated coding sequences, in fact) may play an interesting role since they may be essential for parasite survival.

Second, target validation is essential to reduce the number of potential targets and discard those that are not vital for the parasites. This step may reduce considerably the cost of subsequent experiments, since only those targets with the highest potential usefulness will pass to the next step. RNAi (described above) is probably the most common tool used for this purpose. Expression microarray analysis also allows us to study the expression profile of each gene in the genome at different stages of the life cycle or under different environmental conditions, an important step in identifying genes expressed during the relevant stage of infection and to gain clues as to gene function.

Once target genes have been validated they are tested against different chemicals to identify possible inhibitors. The most successful will be modified for safety and efficacy and then tested *in vitro* and *in vivo* to identify those with the best drug properties.

Although the future is very promising there are still many problems related to the drug discovery process. Target validation is not always reliable since a gene or metabolic pathway may look essential *in vitro* but may not display the same phenotype *in vivo*. Also, it is not always pharmacologically possible to find a 100% inhibitor for the selected target. But the genome project has increased the odds: the possibility of screening thousands of genes increases the probability of finding targets that comply with all the required characteristics to develop a successful drug. Several reviews on drug target discovery using genome sequences in trypanosomatids and other organisms are available (Gutierrez, 2000; Fairlamb, 2003; McCarter, 2004).

Although there is still a long way ahead in the discovery of efficient drugs against trypanosomatids, there is renewed determination. Publication of the *T. brucei*, *T. cruzi* and *L. major* genomes has already changed the way some researchers tackle drug target discovery in these parasites. Comparison of this information with new genome sequences from other species of trypanosomatids such as *T. congolense*, *T. vivax*, *L. braziliensis* and *L. infantum*, currently in the pipeline, will increase the possibilities for drug target discovery for a wider spectrum of trypanosomatids, including those that infect cattle and other livestock. Although the full potential of this massive amount of information is yet unrealized, and the observed variation in pathogenesis is greater than that observed in the sequenced genomes, it is hoped and expected that in the near future research into drug discovery against trypanosomatids will progress more rapidly and eventually lead to the commercialization of effective and affordable drugs for those who need them.

References

Aksoy, S., Williams, S., Chang, S. and Richards, F.F. (1990) SLACS retrotransposon from *Trypanosoma brucei gambiense* is similar to mammalian LINEs. *Nucleic Acids Res* 18 (4): 785–792.

Barry, J.D., Ginger, M.L., Burton, P. and McCulloch, R. (2003) Why are parasite contingency genes often associated with telomeres? *Int J Parasitol* 33 (1): 29–45.

Bastien, P., Blaineau, C. and Pagès, M. (1992) Molecular karyotype analysis in *Leishmania*. In: Avila, J.L. and Harris, J.R. (eds) *Intracellular Parasites*. Plenum Press, New York.

Bastien, P., Blaineau, C., Britto, C., *et al.* (1998) The complete chromosomal organization of the reference strain of the *Leishmania* genome project, *L. major* 'Friedlin'. *Parasitol Today* 14: 301–303.

Berriman, M., Hall, N., Sheader, K., Bringaud, F., Tiwari, B., Isobe, T., Bowman, S., Corton, C., Clark, L., Cross, G.A., *et al.* (2002) The architecture of variant surface glycoprotein gene expression sites in *Trypanosoma brucei*. *Mol Biochem Parasitol* **122** (2): 131–140.

Berriman, M., Ghedin, E., Hertz-Fowler, C., Blandin, G., Renauld, H., Bartholomeu, D.C., Lennard, N.J., Caler, E., Hamlin, N.E., Haas, B. (2005) The genome of the African trypanosome *Trypanosoma brucei*. *Science* **309** (5733): 416–422.

Beverley, S.M. (1996) Hijacking the cell: parasites in the driver's seat. *Cell* **87** (5): 787–789.

Borst, P. (2002) Antigenic variation and allelic exclusion. *Cell* **109** (1): 5–8.

Bringaud, F., Vedrenne, C., Cuvillier, A., Parzy, D., Baltz, D., Tetaud, E., Pays, E., Venegas, J., Merlin, G. and Baltz, T. (1998) Conserved organization of genes in trypanosomatids. *Mol Biochem Parasitol* **94** (2): 249–264.

Bringaud, F., Garcia-Perez, J.L., Heras, S.R., Ghedin, E., El-Sayed, N.M., Andersson, B., Baltz, T. and Lopez, M.C. (2002a) Identification of non-autonomous non-LTR retrotransposons in the genome of *Trypanosoma cruzi*. *Mol Biochem Parasitol* **124** (1–2): 73–78.

Bringaud, F., Biteau, N., Melville, S.E., Hez, S., El-Sayed, N.M., Leech, V., Berriman, M., Hall, N., Donelson, J.E. and Baltz, T. (2002b) A new, expressed multigene family containing a hot spot for insertion of retroelements is associated with polymorphic subtelomeric regions of *Trypanosoma brucei*. *Eukaryotic Cell* **1** (1): 137–151.

Burleigh, B.A. and Andrews, N.W. (1998) Signaling and host cell invasion by *Trypanosoma cruzi*. *Curr Opin Microbiol* **1** (4): 461–465.

Burri, C., Stich, A. and Brun, R. (2004) Current chemotherapy of human African trypanosomiasis. In: Maudlin, I., Holmes, P.H. and Miles, M.A. (eds) *The Trypanosomiases*, CABI Publishing, Wallingford, UK: 403-419.

Caler, E.V., Vaena de Avalos, S., Haynes, P.A., Andrews, N.W. and Burleigh, B.A. (1998) Oligopeptidase B-dependent signaling mediates host cell invasion by *Trypanosoma cruzi*. *EMBO J* **17** (17): 4975–4986.

Callejas, S., Leech, V., Reitter, C. and Melville, S.E. (2006) Hemizygous subtelomeres of an African trypanosome chromosome may accout for over 75% of chromosome length. *Genome Research*, in press.

Cano, M.I., Gruber, A., Vazquez, M., Cortes, A., Levin, M.J., Gonzalez, A., Degrave, W., Rondinelli, E., Zingales, B., Ramirez, J.L. (1995) Molecular karyotype of clone CL Brener chosen for the *Trypanosoma cruzi* genome project. *Mol Biochem Parasitol* **71** (2): 273–278.

Donelson, J.E. (1996) Genome research and evolution in trypanosomes. *Curr Opin Genet Dev* **6** (6): 699–703.

Donelson, J.E., Gardner, M.J. and El-Sayed, N.M. (1999) More surprises from Kinetoplastida. *Proc Natl Acad Sci USA* **96** (6): 2579–2581.

El-Sayed, N.M., Myler, P.J., Blandin, G., Berriman, M., Crabtree, J., Aggarwal, G., Caler, E., Renauld, H., Worthey, E.A., Hertz-Fowler, C., *et al.* (2005a) Comparative genomics of trypanosomatid parasitic protozoa. *Science* **309** (5733): 404–409.

El-Sayed, N.M., Myler, P.J., Bartholomeu, D.C., Nilsson, D., Aggarwal, G., Tran, A.N., Ghedin, E., Worthey, E.A., Delcher, A.L., Blandin, G., *et al.* (2005b)

The genome sequence of *Trypanosoma cruzi*, etiologic agent of Chagas disease. *Science* 309 (5733): 409–415.

Fabre, E., Muller, H., Therizols, P., Lafontaine, I., Dujon, B. and Fairhead, C. (2005) Comparative genomics in hemiascomycete yeasts: evolution of sex, silencing, and subtelomeres. *Mol Biol Evol* 22 (4): 856–873.

Fairlamb, A.H. (2003) Chemotherapy of human African trypanosomiasis: current and future prospects. *Trends Parasitol* 19 (11): 488–494.

Fernandes, A.P., Nelson, K. and Beverley, S.M. (1993) Evolution of nuclear ribosomal RNAs in kinetoplastid protozoa: perspectives on the age and origins of parasitism. *Proc Natl Acad Sci USA* 90 (24): 11608–11612.

Freitas-Junior, L.H., Bottius, E., Pirrit, L.A., Deitsch, K.W., Scheidig, C., Guinet, F., Nehrbass, U., Wellems, T.E. and Scherf, A. (2000) Frequent ectopic recombination of virulence factor genes in telomeric chromosome clusters of *P. falciparum*. *Nature* 407 (6807): 1018–1022.

Fu, G. and Barker, D.C. (1998) Characterisation of *Leishmania* telomeres reveals unusual telomeric repeats and conserved telomere-associated sequence. *Nucleic Acids Res* 26 (9): 2161–2167.

Fu, G., Melville, S., Brewster, S., Warner, J. and Barker, D.C. (1998) Analysis of the genomic organisation of a small chromosome of *Leishmania braziliensis* M2903 reveals two genes encoding GTP-binding proteins, one of which belongs to a new G-protein family and is an antigen. *Gene* 210 (2): 325–333.

Gao, G., Nara, T., Nakajima-Shimada, J. and Aoki, T. (1999) Novel organization and sequences of five genes encoding all six enzymes for de novo pyrimidine biosynthesis in *Trypanosoma cruzi*. *J Mol Biol* 285 (1): 149–161.

Ghedin, E., Bringaud, F., Peterson, J., Myler, P., Berriman, M., Ivens, A., Andersson, B., Bontempi, E., Eisen, J., Angiuoli, S., *et al.* (2004) Gene synteny and evolution of genome architecture in trypanosomatids. *Mol Biochem Parasitol* 134 (2): 183–191.

Gutierrez, J.A. (2000) Genomics: from novel genes to new therapeutics in parasitology. *Int J Parasitol* 30 (3): 247–252.

Hall, N., Berriman, M., Lennard, N.J., Harris, B.R., Hertz-Fowler, C., Bart-Delabesse, E.N., Gerrard, C.S., Atkin, R.J., Barron, A.J., Bowman, S., *et al.* (2003) The DNA sequence of chromosome I of an African trypanosome: gene content, chromosome organisation, recombination and polymorphism. *Nucleic Acids Res* 31 (16): 4864–4873.

Hall, N., Karras, M., Raine, J.D., Carlton, J.M., Kooij, T.W., Berriman, M., Florens, L., Janssen, C.S., Pain, A., Christophides, G.K., *et al.* (2005) A comprehensive survey of the Plasmodium life cycle by genomic, transcriptomic, and proteomic analyses. *Science* 307 (5706): 82–86.

Hargrove, J.W. (2004) Tsetse population dynamics. In: Maudlin, I., Holmes, P. H. and Miles, M. A. (eds) *The Trypanosomiases*, CABI Publishing, Wallingford, UK: 113–137.

Hasan, G., Turner, M.J. and Cordingley, J.S. (1984) Complete nucleotide sequence of an unusual mobile element from *Trypanosoma brucei*. *Cell* 37 (1): 333–341.

Henriksson, J., Porcel, B., Rydaker, M., Ruiz, A., Sabaj, V., Galanti, N., Cazzulo, J.J., Frasch, A.C. and Pettersson, U. (1995) Chromosome specific markers reveal conserved linkage groups in spite of extensive chromosomal size variation in *Trypanosoma cruzi*. *Mol Biochem Parasitol* **73** (1–2): 63–74.

Henriksson, J., Aslund, L. and Pettersson, U. (1996) Karyotype variability in *Trypanosoma cruzi*. *Parasitol Today* **12** (3): 108–114.

Ivens, A.C., Peacock, C.S., Worthey, E.A., Murphy, L., Aggarwal, G., Berriman, M., Sisk, E., Rajandream, M.A., Adlem, E., Aert, R., *et al.* (2005) The genome of the kinetoplastid parasite *Leishmania major*. *Science* **309** (5733): 436–442.

Katinka, M.D., Duprat, S., Cornillot, E., Metenier, G., Thomarat, F., Prensier, G., Barbe, V., Peyretaillade, E., Brottier, P., Wincker, P. (2001) Genome sequence and gene compaction of the eukaryote parasite *Encephalitozoon cuniculi*. *Nature* **414** (6862): 450–453.

Kent, W.J. and Zahler, A.M. (2000) Conservation, regulation, synteny, and introns in a large-scale *C. briggsae–C. elegans* genomic alignment. *Genome Res* **10** (8): 1115–1125.

Kimmel, B.E., ole-MoiYoi, O.K. and Young, J.R. (1987) Ingi, a 5.2-kb dispersed sequence element from *Trypanosoma brucei* that carries half of a smaller mobile element at either end and has homology with mammalian LINEs. *Mol Cell Biol* **7** (4): 1465–1475.

Lopes, E.R. and Chapadeiro, E. (2004) Pathogenesis of American trypanosomiasis. In: Maudlin, I., Holmes, P.H. and Miles, M.A. (eds) *The Trypanosomiases*, CABI Publishing, Wallingford, UK: 303–330.

Martin, F., Maranon, C., Olivares, M., Alonso, C. and Lopez, M.C. (1995) Characterization of a non-long terminal repeat retrotransposon cDNA (L1Tc) from *Trypanosoma cruzi*: homology of the first ORF with the ape family of DNA repair enzymes. *J Mol Biol* **247** (1): 49–59.

McCarter, J.P. (2004) Genomic filtering: an approach to discovering novel antiparasitics. *Trends Parasitol* **20** (10): 462–468.

McDonagh, P.D., Myler, P.J. and Stuart, K. (2000) The unusual gene organization of *Leishmania major* chromosome 1 may reflect novel transcription processes. *Nucleic Acids Res* **28** (14): 2800–2803.

Melville, S.E., Leech, V., Gerrard, C.S., Tait, A. and Blackwell, J.M. (1998) The molecular karyotype of the megabase chromosomes of *Trypanosoma brucei* and the assignment of chromosome markers. *Mol Biochem Parasitol* **94** (2): 155–173.

Melville, S.E., Gerrard, C.S. and Blackwell, J.M. (1999) Multiple causes of size variation in the diploid megabase chromosomes of African trypanosomes. *Chromosome Res* **7** (3): 191–203.

Melville, S.E., Leech, V., Navarro, M. and Cross, G.A. (2000) The molecular karyotype of the megabase chromosomes of *Trypanosoma brucei* stock 427. *Mol Biochem Parasitol* **111** (2): 261–273.

Melville, S.E., Majiwa, P.A.O. and Tait, A. (2004) The African trypanosome genome. In: Maudlin, I., Holmes, P. H. and Miles, M. A. (eds) *The Trypanosomiases*, CABI Publishing, Wallingford, UK: 39–57.

Molyneux, D.H. and Ashford, R.W. (1983) The biology of *Trypanosoma* and *Leishmania*, parasites of man and animals. Taylor and Francis, London.

Moran, M., Ropars, A.L., Guzman, J., Diaz, J. and Garrison, C. (2005) The new landscape of neglected disease drug development [on line]. London School of

Economics and Political Science. The Wellcome Trust. Available from: http://www.wellcome.ac.uk/assets/wtx026592.pdf (Accessed September 2005).

Morel, C.M., Acharya, T., Broun, D., Dangi, A., Elias, C., Ganguly, N.K., Gardner, C.A., Gupta, R.K., Haycock, J., Heher, A.D., *et al.* (2005) Health innovation networks to help developing countries address neglected diseases. *Science* 309 (5733): 401–404.

Myler, P.J., Audleman, L., deVos, T., Hixson, G., Kiser, P., Lemley, C., Magness, C., Rickel, E., Sisk, E., Sunkin, S., *et al.* (1999) *Leishmania major* Friedlin chromosome 1 has an unusual distribution of protein-coding genes. *Proc Natl Acad Sci USA* 96 (6): 2902-2906.

Ngo, H., Tschudi, C., Gull, K. and Ullu, E. (1998) Double-stranded RNA induces mRNA degradation in *Trypanosoma brucei*. *Proc Natl Acad Sci USA* 95 (25): 14687–14692.

Overath, P., Haag, J., Lischke, A. and O'HUigin, C. (2001) The surface structure of trypanosomes in relation to their molecular phylogeny. *Int J Parasitol*, 31 (5-6): 468–471.

Pain, A., Renauld, H., Berriman, M., Murphy, L., Yeats, C.A., Weir, W., Kerhornou, A., Aslett, M., Bishop, R., Bouchier, C., *et al.* (2005) Genome of the host-cell transforming parasite *Theileria annulata* compared with *T. parva*. *Science* 309 (5731): 131–133.

Pays, E., Vanhamme, L. and Perez-Morga, D. (2004) Antigenic variation in *Trypanosoma brucei*: facts, challenges and mysteries. *Curr Opin Microbiol* 7 (4): 369–374.

Pentreath, V. W. and Kennedy, G. E. (2004) Pathogenesis of human African trypanosomiasis. In: Maudlin, I., Holmes, P.H. and Miles, M.A. (eds) *The Trypanosomiases*, CABI Publishing, Wallingford, UK: 283–302.

Porcile, P.E., Santos, M.R., Souza, R.T., Verbisck, N.V., Brandao, A., Urmenyi, T., Silva, R., Rondinelli, E., Lorenzi, H., Levin, M.J., Degrave, W. and Franco da Silveira, J. (2003) A refined molecular karyotype for the reference strain of the *Trypanosoma cruzi* genome project (clone CL Brener) by assignment of chromosome markers. *Gene* 308: 53–65.

Price, H.P., Menon, M.R., Panethymitaki, C., Goulding, D., McKean, P.G. and Smith, D.F. (2003) Myristoyl-CoA:protein N-myristoyltransferase, an essential enzyme and potential drug target in kinetoplastid parasites. *J Biol Chem* 278 (9): 7206–7214.

Rassi, A. and Luquetti, A.O. (2004) Current chemotherapy of American trypanosomiasis. In: Maudlin, I., Holmes, P.H. and Miles, M.A. (eds) *The Trypanosomiases*, CABI Publishing, Wallingford, UK: 421–429.

Ravel, C., Wincker, P., Bastien, P., Blaineau, C. and Pages, M. (1995) A polymorphic minisatellite sequence in the subtelomeric regions of chromosomes I and V in *Leishmania infantum*. *Mol Biochem Parasitol* 74 (1): 31–41.

Robinson, K.A. and Beverley, S.M. (2003) Improvements in transfection efficiency and tests of RNA interference (RNAi) approaches in the protozoan parasite *Leishmania*. *Mol Biochem Parasitol* 128 (2): 217–228.

Rogers, D.J. and Robinson, T. (2004) Tsetse distribution. In: Maudlin, I., Holmes, P.H. and Miles, M.A. (eds) *The Trypanosomiases*, CABI Publishing, Wallingford, UK: 139–179.

Roper, J.R., Guther, M.L., Milne, K.G. and Ferguson, M.A. (2002) Galactose metabolism is essential for the African sleeping sickness parasite *Trypanosoma brucei*. *Proc Natl Acad Sci USA* **99 (9):** 5884–5889.

Roth, C., Bringaud, F., Layden, R.E., Baltz, T. and Eisen, H. (1989) Active late-appearing variable surface antigen genes in *Trypanosoma equiperdum* are constructed entirely from pseudogenes. *Proc Natl Acad Sci USA* **86 (23):** 9375–9379.

Santos, M.R., Cano, M.I., Schijman, A., Lorenzi, H., Vazquez, M., Levin, M.J., Ramirez, J.L., Brandao, A., Degrave, W.M. and da Silveira, J.F. (1997) The *Trypanosoma cruzi* genome project: nuclear karyotype and gene mapping of clone CL Brener. *Mem Inst Oswaldo Cruz* **92 (6):** 821–828.

Scholler, J.K., Reed, S.G. and Stuart, K. (1986) Molecular karyotype of species and subspecies of *Leishmania*. *Mol Biochem Parasitol* **20 (3):** 279–293.

Shapiro, T.A. and Englund, P.T. (1995) The structure and replication of kinetoplast DNA. *Annu Rev Microbiol* **49:** 117–143.

Shlomai, J. (2004) The structure and replication of kinetoplast DNA. *Curr Mol Med*, **4 (6):** 623–647.

Smith, T.K., Crossman, A., Brimacombe, J.S. and Ferguson, M.A. (2004) Chemical validation of GPI biosynthesis as a drug target against African sleeping sickness. *EMBO J* **23 (23):** 4701–4708.

Stevens, J.R. and Gibson, W.C. (1999) The evolution of pathogenic trypanosomes. *Cad Saude Publica* **15 (4):** 673–684.

Stevens, J.R., Noyes, H.A., Schofield, C.J. and Gibson, W. (2001) The molecular evolution of *Trypanosomatidae*. *Adv Parasitol* **48:** 1–56.

Steverding, D., Stierhof, Y.D., Chaudhri, M., Ligtenberg, M., Schell, D., Beck-Sickinger, A.G. and Overath, P. (1994) ESAG 6 and 7 products of *Trypanosoma brucei* form a transferrin binding protein complex. *Eur J Cell Biol* **64 (1):** 78–87.

Sunkin, S.M., Kiser, P., Myler, P.J. and Stuart, K. (2000) The size difference between *Leishmania major* friedlin chromosome one homologues is localized to sub-telomeric repeats at one chromosomal end. *Mol Biochem Parasitol* **109 (1):** 1–15.

The Wellcome Trust (2005). Drug discovery at Dundee [on line]. The Wellcome Trust. Available from: http://www.wellcome.ac.uk/doc_wtx027342.html (Accessed 25 October 2005).

Van der Ploeg, L.H., Schwartz, D.C., Cantor, C.R. and Borst, P. (1984) Antigenic variation in *Trypanosoma brucei* analyzed by electrophoretic separation of chromosome-sized DNA molecules. *Cell* **37 (1):** 77–84.

Villanueva, M.S., Williams, S.P., Beard, C.B., Richards, F.F. and Aksoy, S. (1991) A new member of a family of site-specific retrotransposons is present in the spliced leader RNA genes of *Trypanosoma cruzi*. *Mol Cell Biol* **11 (12):** 6139–6148.

WHO (2002) *The World Health Report*. World Health Organization, Geneva.

Wickstead, B., Ersfeld, K. and Gull, K. (2004) The small chromosomes of *Trypanosoma brucei* involved in antigenic variation are constructed around repetitive palindromes. *Genome Res* **14 (6):** 1014–1024.

Wincker, P., Ravel, C., Blaineau, C., Pages, M., Jauffret, Y., Dedet, J.P. and Bastien, P. (1996) The *Leishmania* genome comprises 36 chromosomes conserved across widely divergent human pathogenic species. *Nucleic Acids Res* **24 (9):** 1688–1694.

Zhang, J. (2000) Protein-length distributions for the three domains of life. *Trends Genet*, **16 (3):** 107–109.

2

The practical implications of comparative kinetoplastid genomics

C. S. Peacock

1 Introduction

The kinetoplastids are a remarkable group of organisms, quite unlike any other. All are unicellular protozoan organisms that represent one of the earliest branches of the eukaryotes. They have many unique and extraordinary characteristics and can tell us much about the origin of the eukaryotic lineage. They are also one of the least understood of all orders. Until recently, they were one of the few major groups that had no representative sequenced genome in the public databases. And yet these are not just model organisms, interest in them is not restricted to pure research into understanding their niche in the animal kingdom or the evolution of eukaryotic biological systems. There is a far greater reason to understand more about these organisms. Although the order is large, with over 500 known members across 22 genera, a number of them are important human and veterinary pathogens. The three main human pathogens within the kinetoplastida cause many thousands of deaths each year and serious illness to millions. Some 500 million people, a fifth of the world's population, are at risk (www.who.int/tdr/). The three diseases that they cause in humans are more commonly known as leishmaniasis, sleeping sickness and Chagas disease. Unlike bacterial diseases, the methods for dealing with these devastating infections have not significantly changed for more than half a century. There are no vaccines, no prophylaxis and very little in terms of new drug treatments.

This review will highlight the current problems in treating these diseases and look at the impact the recent completion of the genome sequences for *Leishmania major*, *Trypanosoma brucei* and *Trypanosoma cruzi* has had on research. It will also look at how the availability of the genome sequences has stimulated large-scale post-genomic studies and, importantly, what we can expect in return from the investment of sequencing these genomes.

The order Kinetoplastida can be segregated into two suborders, Bodonidae and Trypanosomatidae, based both on physiology and genetic phylogeny (Simpson *et al.*, 2004). Members of this order have certain defining morphological features in common.

Comparative Genomics and Proteomics in Drug Discovery, edited by John Parrington and Kevin Coward. © 2007 Taylor and Francis Group.

Visible at the light microscopic level, they have a unique and characteristic kinetoplast from which they derive their name. This is a specialised independently replicating organelle that lies within the single large specialised mitochondrion and is associated with the flagella basal body. Belonging to the phyla Euglenozoa, all are unicellular with long whip-like flagellum, and many of the genera are free living or commensal. However, this order contains important pathogens, many of which fall into the sub-order Trypanosomatidae. All members of this group are obligate pathogens and parasitise a wide range of organisms. This review is restricted to those few that are of major medical importance. Members of the kinetoplastids causing human disease are restricted to approximately 15 species of leishmania, 2 species of African trypanosomes and a single South American trypanosome called *Trypanosoma cruzi*. As an order, the kinetoplastids are also important pathogens of domestic animals and thus their impact on human activity goes beyond the devastating effect they have on human health. African trypanosomes kill millions of cattle each year causing further economic decline in some of the poorest countries in the world. *Leishmania* spp. infect many types of mammalian hosts including the domestic dog. This not only presents as a veterinary problem but the close proximity to humans means this reservoir of infection has an impact on human disease. Kinetoplastid parasites of economic relevance outside of those described in this review include many species of the Bodonids, which are important pathogens of fish, insects and plants.

Although there are almost a dozen whole genome projects and sporadic genomic data in the public databases for as many species again, only three of these have been completed and published and as such will form the basis of this chapter. As in the recently published genome papers, these three, *Leishmania major*, *Trypanosoma brucei* and *Trypanosoma chagas*, will be referred to as Tritryps (Berriman *et al.*, 2005; El-Sayed *et al.*, 2005a, 2005b; Ivens *et al.*, 2005). The other genome projects will be mentioned and summarised with respect to future prospects within this field.

2 The diseases

Leishmaniasis is not a single entity but describes a very broad spectrum of disease that can be broadly divided into three clinical groups (Dedet and Pratlong, 2003). The commonest is cutaneous leishmaniasis, resulting in one or more ulcerating skin lesions that take months or even years to heal, are prone to secondary infection and lead to permanent scarring. The scarring is so severe that to prevent facial disfigurement certain cultures have practiced 'leishmanisation' for hundreds of years, involving the inoculation of infective material from a patient's ulcer into the healthy tissue in a hidden site, inducing a cutaneous lesion that 'protects the recipient from further lesions on exposed areas of the body' (Sukumaran and Madhubala, 2004). At the other end of the spectrum is visceral leishmaniasis, a systemic progressive disease of the liver, spleen and bone marrow that results in almost certain death without vigorous treatment and, with the exception of malaria, kills more people than any other parasitic disease (http://www.who.int/tdr/diseases/default.htm). Mucocutaneous leishmaniasis is a progressive chronic destruction of the mucosal layer, usually involving the soft palate and nasal pharyngeal tissue. Although human genetic variability influences disease progression (Blackwell *et al.*, 2004), the type of disease is primarily dependent on the infecting species of leishmania. There are 15 species that regularly infect humans and a handful of others that are rarely diagnosed in humans. African trypanosomiasis

Table 1. Overview of the Tritryp diseases.

	Leishmania species	African trypanosomes	*Trypanosoma cruzi*
Disease	Visceral leishmaniasis Cutaneous leishmaniasis Mucocutaneous leishmaniasis	Sleeping sickness	Chagas disease
Incidence/infected	2 million/12 million	60,000/300,000	16 million
At risk	350 million		120 million
Deaths	59,000	48,000	14,000
DALYs	2.36 million	1.59 million	667,000
Vector	Sandfly	Tsetse fly	Triatomine bugs
Host infection	Intra-cellular in macrophages	Extra-cellular in blood	Intra-cellular in range of host cells
Other vertebrate hosts	Dogs, rodents, wild mammals	Cattle (Ngana)	
Current epidemiology	Incidence and geographical spread increasing	Almost eradicated in 1960s, incidence increasing now	Incidence decreasing
Geography	88 countries, South America, Africa, southern Europe, Asia	36 countries across Africa	18 countries in South and Central America
Treatment problems	Widespread drug resistance, toxicity	Drug toxicity, poor response in second-stage disease	Poor response with chronic disease
Other	Opportunistic infection in HIV		

causes a progressive disease that, untreated, leads invariably to neurological infection, coma and death (Burri and Brun, 2003). Two recognised species infect humans, *Trypanosoma brucei gambiensie* in west and central Africa and *Trypanosoma brucei rhodesiense* in east and southern Africa. Other species cause fatal disease in cattle. Infection with *Trypanosoma cruzi*, the causative agent of Chagas disease, can lead to a number of clinical presentations, from a mild, short-lived, acute fever, acute myocarditis through to the classic chronic pathology that develops many years later. Symptoms associated with chronic illness include myocardiopathy, aneurysms and enlargement of the heart and intestine (Miles, 2003).

3 Current treatment

At present, the situation for a vast majority of patients with one of these diseases is often bleak. These diseases are endemic in the poorest countries in the world where the majority of patients have an income of less than $2 a day. Even apart from the fact that most of those affected are from developing countries with health systems that are inadequately equipped, there are no licensed vaccines for any of these diseases and no prophylactic treatments available. The situation is further compounded by the lack of investment in drug treatment, such that front line drugs for these diseases have not changed for decades (Croft et al., 2005). This has led to the situation on the ground where drugs, when available, are toxic, and difficult to administer. A further complication of long-term use combined with poor treatment regimes is the emergence of widespread drug resistance. Unlike many bacterial diseases, the paucity of available drugs means that multidrug therapy, used to avoid the development of drug resistance cannot be applied. Melarsoprol, the only arsenical drug still in use, is used to treat second-stage sleeping sickness, a treatment regime so drastic that the treatment alone kills 2–12% of patients (Pepin and Milord, 1991). Given that exposure to leishmania imparts a long lasting or even permanent immunity to the development of future disease, it had optimistically been expected that vaccination would be a plausible and relatively inexpensive way of protecting those at risk from infection. However, despite many years of trying and some success in animal models, there is as yet no human vaccine available. Immunity appears to be maintained by persistent parasites surviving in the host (Solbach and Laskay, 2000), evidenced by the recurrence of disease following cell-mediated immune suppression and the emergence of latent leishmaniasis associated with HIV.

Currently, apart from barrier methods to prevent infected insect vectors from biting and passing on the pathogen to humans, the only means of combating these devastating diseases is by treating patients after they present to health workers.

4 The moral issue

So why, given the massive advances in medical treatments over the last few decades, are new drugs and treatments so needed for these diseases not forthcoming? The need for new treatments for all of these diseases is demonstrated by their declaration as 'Neglected diseases' by the World Health Organization (WHO), Médecins Sans Frontières (MSF) and the Gates Foundation. The WHO has prioritised ten tropical diseases for research and investment in the next decade (http:www.who.int/tdr/). Together with malaria these three diseases make up the four protozoan pathogens on the list. Leishmaniasis and sleeping sickness feature in the list of six prominent diseases to be targeted by MSF (http://www.accessmed-msf.org/). In 2003, seven institutions, including MSF and the WHO formed the Drugs for Neglected Diseases Initiative to try and raise the awareness of neglected diseases affecting the developing world and to try and coordinate drug development (http://www.dndi.org). They currently have only four diseases in their portfolio, three of which are leishmaniasis, sleeping sickness and Chagas disease. The Bill and Melinda Gates Foundation has all of these diseases on its list for funding and recently gave $10 million for research into vaccines for leishmaniasis and $30 million to the Institute for OneWorld Health to help reduce the mortality and morbidity due to visceral leishmaniasis. It has also awarded a further $15 million for research into drug development for leishmaniasis and sleeping sickness and $4.6 million for drug development in leishmaniasis and Chagas disease (www.gatesfoundation.org/default.htm).

The major obstacle to developing new drugs or even providing access to current drugs is financial. These diseases predominantly affect those people living in the tropics and sub-tropics, including many of the poorest countries on the planet. This 'Third World market' does not generate sufficient income to interest drug companies; as a result, the main treatments for leishmaniasis and African trypanosomiasis are over 50 years old and toxic to the patient. Recently this has brought financial and moral issues into conflict. Two examples dramatically illustrate this point. There are only two recognised drugs available to treat early stages of Chagas disease and no effective treatment for the chronic stage at which most cases are identified. Nifurtimox, the less toxic of the two, was threatened with withdrawal in 2001 but three years later the drug company Bayer agreed to donate a year's supply to WHO. Similarly, the drug company Roche signed over the rights and technology to the Brazilian government to produce the drug benzonidazole, the only alternative treatment for Chagas disease. The second example was in the treatment of African trypanosomiasis. In 1995, the drug eflornithine, used to treat late-stage sleeping sickness caused by *Trypanosoma gambiense,* was withdrawn by the drug company Aventis. The only other drug to treat the disease at this stage is the arsenical melarsoprol whose long-term use has led to reduced efficacy and resistance in recent epidemics (Brun *et al.*, 2001). The active ingredient of eflornithine was, however, licensed to Bristol-Myers Squibb's (BMS) to produce Vaniqa, a cream to remove female facial hair. In 2001, a change in policy resulted in Aventis supplying MSF with the drugs eflornithine, melarsoprol and pentamidine and financial support for the WHO's initiative for drug development in sleeping sickness (http://www.who.int/tdr/research/progress9900/tools/drug-tryps.htm).

Although the main reason for neglect is financial, there is a significant lack of understanding in kinetoplastid biology. The mechanism by which most of the drugs currently in use effect their antiprotozoal activity is unknown (*Table 2*). For example, the main treatment for visceral leishmaniasis for the last 60 years has been various formulations of pentavalent antimonials and yet the mode action of these frontline drugs is still unclear (Croft *et al.*, 2006).

One of the most attractive methods of drug development is to take a compound that has already been used or trialled for another disease. This not only shortens the time to get the drug to production and distribution but also dramatically reduces development costs. A case in point is the recent exciting development and licensing of the orally delivered drug miltefosine for visceral leishmaniasis. All other affordable chemotherapeutic treatments for this disease rely on daily intra-muscular injections of a drug for a month, something that stretches already fragile health systems. Miltefosine was originally developed for use in cancer treatment (Eibl and Unger, 1990), meaning that toxicity tests on animals and humans had already been performed. However, its mode of action has not been completely determined although it is known to be a phospholipid analogue. Clinical trials in India have shown miltefosine to be very effective against visceral leishmaniasis and it has recently been licensed for use in India and South America. Knowledge of the complete biology and metabolic pathways of these pathogens would invariably speed up this process of validating existing licensed compounds for use against these diseases and present exciting new opportunities to target novel pathogen proteins.

5 The Tritryp genome projects

Comparative genomics of the complete genomes of related pathogens is proving to be a useful methodology for generating new targets for drugs and vaccines. The large size of the eukaryotic genomes has meant that, until recently, this was restricted solely to bacterial diseases. In recent years, advances in sequencing technology and computer power have led to much larger genome projects being undertaken.

The initial idea to sequence the three Tritryp genomes was made at the WHO-TDR-funded Parasite Network Planning Meeting in Rio de Janeiro in 1994. This meeting was attended by 40 scientists from across the globe and began the era of collaborative research that would epitomise the future sequencing efforts. The purpose

Table 2. *Drugs currently used for treating leishmaniasis, sleeping sickness and Chagas disease.*

Drug name	Disease	Dose/regime	Target	Year	Problems
Pentamidine isethionate	Sleeping sickness	7–10 injections/daily 6 injections over a month		1950	only early stage/ often fails in T. rhodiense/toxic
Suramin	Sleeping sickness		non-specific enzyme inhibitor	1922	early stage only/toxic
Berenil	Sleeping sickness	injection 9–12 injections over a month	intercalate with DNA		Unlicensed, mainly used in cattle, toxic in humans
Malarsoprol	Sleeping sickness		unknown	1949	arsenical, very toxic
Eflornithine	Sleeping sickness	4 infusions/day for 2 weeks	ornithine decarboxylase inhibitor	1990	Only T. gambiense/toxic
Nifurtimox	Chagas/sleeping sickness	oral/3 doses/day for 3 months	unknown	1960	poor efficacy in chronic cases
Benznidazole	Chagas	oral/2 doses/day for 2 months	unknown	1974	poor efficacy in chronic cases
Pentavalent antimonials	All leishmaniasis	injection/daily/1 month	unknown	1920	toxic/drug resistance
Amphotericin B	VL/MCL	14–20 infusions	ergosterol-like sterols	1955	toxic
Liposomal AmphotericinB	VL/VL in HIV	4 i.v injections	blocks thymidine synthase	1962	toxic/very expensive
Miltefosine	VL	oral/1 dose/day 1 month	phosphocholine analogue	2002	
Aminosidine (Paromomycin)	CL/VL	injection/daily/10 days	unknown, protein inhibition	1999	no drug partner
Interferon gamma	VL/DCL/MCL	injection/combination treatment	upregulates macrophage activation		very expensive/needs combination therapy
Imidazoles	CL	oral	inhibit sterol synthesis		low efficacy better for canine leishmaniasis,
Allopurinol	CL/VL	oral	purine analogue	1980s	poor in humans
Sitamaquine	VL	oral	unknown	2005	phase II trials

of the meeting was to select the organisms to be worked on and ultimately sequenced, to establish 'genome networks' for each of these and to decide on common protocols, data storage facilities and software. At that time, current technology was inadequate for the volume of sequence that would be necessary to complete a chromosome yet alone a whole genome. However, it was predicted that there would be massive improvements to sequencing technology, which would lead ultimately to finished and, more importantly, publicly available complete genomes. Furthermore, it was recognised that despite the inequality of funding between research in the endemic countries and that in Europe and America, every attempt should be made to 'share' the workload and include institutions in South America and Africa.

In the group Kinetoplastida, three species, chosen to represent the main pathogens, were selected for sequencing. As well as the medical and veterinary importance of these species, the other criteria for selecting these particular species were how well they could be grown and maintained in the laboratory and how much prior research had been conducted on it. To represent the leishmaniasis, *Leishmania major* MHOM/IL/81/Friedlin strain was chosen based on its proven pathogenicity and ease of maintenance in the laboratory. African trypanosomes were represented by the *Trypanosoma brucei brucei* TREU927/4 GUTat10.1 strain based on prior work on genetic mapping and ease of maintenance in the laboratory and the *Trypanosoma cruzi* strain CL Brenner was similarly chosen for the depth of previous experimental characterisation. Each of the projects proceeded in the manner determined by the respective members of the consortium. The methodologies were determined by the available technologies. Initial pilot projects to determine the feasibility of the idea were based on sequencing through small BAC clones and on EST and GSS projects to attempt some sort of whole genome approach. The advent of high-throughput technologies in the late 1990s led to more ambitious projects and the development of sequencing consortia that realistically had both the infrastructure and funding to attempt sequencing the whole genome. The initial work involved physical mapping of the genomes for each organism. *L. major* was shown to have 36 chromosomes (Wincker *et al.*, 1996), *T. brucei* 11 very large (mega base) chromosomes and a number of intermediate and mini chromosomes (Turner *et al.*, 1997). The karyotype of the Brenner strain of *T. cruzi* proved impossible to elucidate, due to the repetitive and very variable nature of the homologues of the genome.

The period of time over which these projects have evolved, the different sequencing centres involved, and the organism-specific problems faced in mapping and sequencing led initially to separate sequencing consortia and different sequencing strategies. *T. brucei* was half sequenced using a BAC-by-BAC approach and half by whole chromosome shotgun. The *L. major* genome was initially started as a cosmid sequencing project, but a majority of it (~80%) was carried out by separating the chromosomes using pulse field gel electrophoresis (PFGE) and using a whole chromosome shotgun approach. The inability to produce an accurate physical map of the *T. cruzi* strain and the variability of the homologues meant that the initial BAC-by-BAC project made way for a whole genome shotgun approach followed by contig assembly. In 1998 the smallest chromosome from *L. major* (Myler *et al.*, 1999) became the first kinetoplastid chromosome to be fully sequenced and published. For the first time, some of the unusual features of genome architecture in the kinetoplastids were revealed.

These three pathogens have many physical and behavioural similarities. They all undergo morphological changes in both the mammalian host and vector. The necessity for a period of development in an insect vector in which there is both a rapid clonal expansion and morphological alteration from a non-infective promastigote form to an infective metacyclic promastigote specifically adapted to survive within the mammalian host is also common to all three. They each have a very effective impermeable protective cell surface, exchange nutrients using a specialised flagella pocket and share an unusual array of subcellular structures such as the kinetoplast, glycosomes, subpellicular microtubules, paraflagellar rod and acidocalcisomes. However, despite their common ancestry, their differences are equally as striking, particularly in their interaction with the mammalian host. *T. brucei* is an extracellular parasite residing mainly in the blood. *T. cruzi* and *Leishmania* spp. invade host cells and remain intracellular, although the former reside in the cell cytoplasm whereas *Leishmania* resides in the phagolysosome of macrophages, the very cells that are designed to kill invading microbes and probably the most hostile environment possible. Adaptation to survival in different environments in the host has led to development of strikingly different ways to evade the immune system. *Leishmania* alters macrophage activation and suppresses T cell-mediated immunity (Teixeira *et al.*, 2006). *T. cruzi* also interferes with the response of the host to invasion (Andrade and Andrews, 2005) and *T. brucei* evades complement and antibody by periodic switching of the variable surface glycoprotein (VSG) that covers its outer membrane (Donelson, 2003). Each of the Tritryps also has a rigid requirement for a specific vector, a feature that determines their geographical spread.

The growing realisation that despite the vast differences in pathology and life style, these three pathogens showed a remarkable similarity at the genome level, led to the three sequencing projects being loosely coordinated under a single Tritryp organisation comprising of four sequencing centres (the Wellcome Trust Sanger Institute, the Institute for Genomic Research, the Seattle Biomedical Research Institute and the Karolinska Institute) and the extensive scientific community. In 2005, a full 10 years after the initial parasite genome network meeting and 7 years after the first sequencing, all three projects were finished and the three complete genomes were published together (Berriman *et al.*, 2005; El-Sayed *et al.*, 2005a; Ivens *et al.*, 2005). The complete sequences, together with the in-silico prediction, annotation and curation of published data by a group of more than 200 researchers from 26 countries, represents a truly global effort to initiate the benefits of the sequencing initiatives.

6 Tritryp genome architecture

As one of the earliest branches of the eukaryotes, it was expected that there would be some specific differences between the kinetoplastids and the higher eukaryotes that harbour these parasites. The scale of the physical differences in genome architecture and variety of unusual and unique functional systems uncharacteristic of other eukaryotes was far greater than anticipated.

The organisation of the genes in this group is so far unique, with all the genes being organised in clusters on a single strand of the DNA (with the exception of the specialised variable surface antigens of *T. brucei*). A directional gene cluster (DGC) can represent anything from a single gene to more than 100. At the end of the DGC the genes switch to the opposite strand. Instead of each gene having a promoter, it is likely that polycistronic transcription is initiated bi-directionally at the strand switch region between divergent DGCs and terminates at the convergent strand switch region (Martinez-Calvillo *et al.*, 2004). Instead of being transcribed separately the genes on a single DGC are transcribed as single unit with the genes being trans-spliced and extensively processed after transcription. Each stable mRNA transcript is derived by excising the coding sequence and adding a 5′ capped spliced leader sequence and polyadenylating the 3′ end. Another distinctive feature of transcription in these organisms is that cis-splicing is almost completely absent. Only four genes out of more than 8000 were found to have a single intron, all of which appear to be RNA binding proteins. The spliceosome appears capable therefore of catalysing both cis- and trans-splicing (Liang *et al.*, 2003). Unique to this order, this polycistronic method of transcribing genes means that regulation of protein expression is not controlled at the level of gene transcription in the same way as other eukaryotes. Instead gene expression levels are controlled in two other ways. First, post-transcriptional modification of transcripts is very important, a reflection of which is the large number of genes with predicted RNA recognition motifs. Second, another aspect that appears to be unique to this group is occurrence of gene duplication into multiple tandem repeat units of identical genes. Duplication events in other eukaryotes that result in identical copies of a gene being created will lead either to a change in function of the second copy or to its degeneration into a non-functional pseudogene. Kinetoplastids on the other hand usually retain the integrity of duplicate copies, to the extent that some genes occur in large extended multigene arrays of functional identical genes.

Genome plasticity is a feature of the Tritryps, and although common to all three, it is expressed in quite different ways. In *T. cruzi* there is a massive amount of duplication, so much so that more than 50% of the genome is repeated sequence, made up of both non-coding repeated sequences such as retrotransposons and large gene families. Most of these repeated gene families express surface molecules involved in host interactions. The largest family, the mucin-associated surface proteins, has over 1300 members (El-Sayed *et al.*, 2005b). In *T. brucei* there are large subtelomeric arrays that contain hundreds of variant surface glycoprotein genes (VSGs). There are at least 1000 of these VSG genes which are not expressed and occur as functional genes, full length pseudogenes and incomplete gene fragments. Expression requires an intact gene to be inserted into a specific subtelomeric bloodstream from expression site, which are associated with co-transcribed expression site-associated genes. These VSG genes are clonally activated on infection in the host and cover the surface of the parasite in a dense coat, shielding it from the host's immune recognition system. Occasionally, VSG rearrangement occurs at the expression site, covering the parasite in a new antigenically different coat, thereby allowing a proportion of the parasites to always avoid the humoral immune response. The vast number of intact, pseudogene and remnant genes are thought to be so mobile that they provide an inexhaustible repertoire of new VSGs (Donelson, 2003). It is because of this very specialised immune avoidance system of *T. brucei* that, unlike leishmaniasis, developing a vaccine against the human infective form is considered a bleak prospect. These large gene families tend to be associated with the parasite's co-evolution with its host and as such tend to be genus, species and even strain specific. The subtelomeres in both *T. brucei* and *T. cruzi* are prone to a great deal of insertions/deletions and duplications. Even homologues of *T. brucei* chromosomes differ significantly in sequence and size and as such different isolates can vary in genome content by as much as 25% (Melville *et al.*, 1998). *L. major* does not have such dramatic rearrangement in the subtelomeres; in fact these regions are short and while they do contain genus-specific genes in some cases they are very well conserved. *L. major* also differs in that it lacks any recognisable active retrotransposable elements, something which might explain the relatively stable nature of the chromosome structure. One area of genome plasticity *Leishmania* does exhibit is an ability to generate extra copies of some chromosomes. Three of the chromosomes appeared tri- or tetra-somic in the genome project. This flexibility in being partially aneuploid has proven detrimental to researchers in the past who have generated double gene knockout experiments only to find the parasite has generated two new copies of the chromosome with the wild type gene intact (Cruz *et al.*, 1993). How and under what circumstances the parasite is able to do this and whether it can be applied to all chromosomes is not understood. It is presumably a novel means of maintaining genome integrity.

7 Core proteome

While each genome has very distinctive genome architecture at the ends of chromosomes, the rest of the genome is highly conserved. The haploid genome of these organisms range from approximately 8300 for *L. major* to 12,000 for *T. cruzi* (*Table 3*). The disparity in numbers between them is mainly accounted for by the expanded gene families in the subtelomeres of the two trypanosomes. The core kinetoplastid proteome is represented by 6158 orthologous genes that occur in all three genomes (El-Sayed *et al.*, 2005b). Remarkably, 5812 of these are found in the same gene order in all three genomes. Thus synteny is conserved for 94% of the genes that form the core proteome. Breaks in synteny tend to occur either at strand switch regions between directional gene clusters, areas of expanded gene families (tandem gene arrays), and positions of retroelements or structural RNA genes. Gene order within directional gene clusters is conserved with insertions, deletions and rearrangements occurring at telomeres, strand switch regions and retroelement sites. Examination of differences between genes within the core proteome and those orthologues found in

Table 3. *Genome statistics for the Tritryps.*

	Leishmania major	*Trypanosoma brucei*	*Trypanosoma cruzi*
Chromosomes	36	11 (plus ~ 100 mini and intermediate chromosomes)	~28
Haploid size	33.6 Mb	26 Mb	55 Mb
Sequencing method	BAC/WCS	BAC/WCS	WGS
Genes per haploid chromosome	8311	9068	~12,000
Species-specific genes	910	1392	3736

The three genomes have considerable difference in overall genome content (although the mini and intermediate chromosomes in *T. brucei* would add another 10 Mb) and to some extent this is reflected in the number of genes.

higher eukaryotes will attract attention as potential drug targets. Unfortunately, approximately half the genes that appear to define the kinetoplastids have no known function. The genome projects will, however, have provided the full complement of genes from which future expression analysis and functional assays can be performed to help characterise these proteins. One of the most interesting potential targets for drug design are those genes that show greatest similarity to prokaryotic sequences. These genes will be amongst those that are most divergent from higher eukaryotes and thus disruption by chemotherapy will have a reduced chance of being toxic to the host. It is probable that some of these may well have been acquired from a bacteria in the past. After rigorous similarity searches and phylogenetic analysis almost 50 genes involved in metabolic processes were identified as having been likely to have been acquired by horizontal transfer from bacteria (Berriman et al., 2005).

8 Species specificity

Domain analysis reveals a similar story of conservation of a basic kinetoplastid pro-teome. Fewer than 5% of domains are species specific, with *L. major* having a little less than 4%, *T. brucei* 1% and *T. cruzi* 2% of their domains absent in the other two species. These species-specific genes and domains reveal some interesting acquisitions by the different pathogens since their divergence from the ancestral kinetoplastid and reflect their co-evolution with the host. For example, *L. major* has two genes that have macrophage migration inhibitory factor (MIF) domains. The macrophage migration inhibitory factor proteins are one of the most pleiotropic cytokines and are involved in several pathways influencing macrophage activation (Donn and Ray, 2004). The similarity to MIF genes in humans raises the possibility that the *L. major* orthologues may inhibit macrophage activation, alter its effectiveness or prevent macrophage lysis after infection. Apart from the domains associated with VSG expression, *T. brucei* has an AOX domain containing protein which acts as an alternative terminal oxidase in mitochondria. Interestingly, *T. cruzi* has several hormone-type domains that are not found in the other two. It is interesting to hypothesise that proteins containing such domains might be involved in some of the reactive changes that occur in Chagas dis-ease. Severe chronic disease is typified by gross pathological changes to the heart and intestine and autoimmune problems associated with pathology (Girones *et al.*, 2005). It is known that a number of *T. cruzi* antigens elicit the production of human anti-bodies that cross react with host tissues and are implicated in heart and renal pathol-ogy (Arce-Fonseca *et al.*, 2005; Matsumoto *et al.*, 2006).

Genes present in all three genomes that appear to be absent or poorly conserved in higher eukaryotes will make good potential drug targets for a wide range of chemotherapeutic agents against all three diseases. A majority of species-specific proteins have evolved in response to co-evolution with the host and as such are good for both chemotherapy and as potential vaccine candidates.

9 Metabolism

Analysis of the complete gene complement of these pathogens will help identify the metabolic pathways that are always an intense area of interest for drug design. Part of the extensive analysis involved in annotating genes using Enzyme Commission (EC) numbers. A novel way of assigning EC numbers uses a program called Gotcha (Martin *et al.* 2004) which uses similarity searches to automatically assign gene ontology (GO) terms to all proteins. These GO terms can be converted to EC numbers using the GO term to EC number mapping file available at the GO consortium website (www.geneontology.org). The EC numbers can then be mapped onto the metabolism pathways at KEGG (Kyoto Encyclopedia of Genes and Genomes -http://www.genome.ad.jp/kegg/pathway.html) and the Tritryp metabolic pathways reconstructed. The advantage of having the whole genome is that the search for enzymes missing in the reconstructed pathway is not then restricted by the limitations of missing enzyme sequences normally imposed on mining incomplete genome data. Without the complete gene complement, an incomplete pathway could be due to divergence of the gene, loss of function in the pathway or absence of the sequence in the database. Searching the complete gene complement allows for less well conserved enzymes to be found. Enzymes on the reconstructed pathways are colour coded to reflect the level of confidence in the correct EC number assignment. As well as those for individual genomes, combined Tritryp pathways were also determined highlighting the metabolic differences between the pathogens. These Tritryp metabolism pathways are available on the web at http://tbdb/bioinformatics.dundee.ac.uk/kegg/. Again, most of the species-specific enzymes could be linked to the different environments that the pathogens encounter in the host and vector. Thus *L. major* is the only one capable of hydrolysing disaccharides, reflecting the fact that sandflies, the vectors of *Leishmania* spp., are the only vectors that are able to feed on nectar and aphid honeydew, providing the infecting promastigotes with an alternative energy source. The range of sugars in the mammalian blood is restricted compared to the intracellular environment, which is reflected in the absence of certain sugar kinases in *T. brucei*. Part of the attachment system of mucins, found in *T. cruzi* as protective surface molecules, utilises aminoethylphosphonate (AEP), the complete synthesis of which is only possible in *T. cruzi*.

There are a number of metabolic pathways that exist in all the kinetoplastids that have long been recognised as fertile ground for searching for effective drug targets (Wang, 1995). A prime example is thiol metabolism in which the kinetoplastids have a unique system: the general eukaryotic system of glutathione and glutathione reductase is replaced with trypanothione and the flavoenzyme trypanothione reductase (Fairlamb *et al.*, 1985). As well as trypanothione, they also contain glutathione, monoglutathionylspermidine and ovothiol, all of which are kept in their reduced state by trypanothione reductase. This enzyme, essential for all the kinetoplasts, is absent in mammals and so drugs that specifically target it are less likely to be toxic to the host. Carbohydrate metabolism is another area where the kinetoplastids are unique. Unlike other eukaryotes, the enzymes for glucose metabolism are physically separated from the rest of the cell in structures called glycosomes (Michels *et al.*, 2000). While this in itself may not be significant, the glycolytic enzymes have significant sequence and structural differences that make them good targets (Berriman *et al.*, 2005). The surface molecules that cover the parasites and protect against physical attack from host complement and immune complexes are attached to the parasite surface using glycosylphosphatidylinositol (GPI) anchors. Correct assimilation of these parasite-specific molecules is essential for survival and infectivity and are therefore a good area for the development of chemotherapeutic agents (Ferguson *et al.*, 1999). Synthesis of the lipid components of parasite membranes is also different to that in mammals, with the kinetoplastids having similar pathways for sterol biosynthesis to fungi. This has prompted research into known anti-fungal sterol inhibitors for their anti-kinetoplastid activity. Drugs targeting fatty acid biosynthesis are being examined for their potential as chemotherapeutic agents against a whole host of bacterial and protozoan diseases. Two known inhibitors of this pathway, thiolactomycin and triclosan, have both been shown to be effective for killing trypanosomes (Morita *et al.*, 2000; Paul *et al.*, 2004).

To retain the integrity of the protective membrane, all trypanosomes restrict endocytosis of molecules into the cell to the specialised flagella pocket. Blocking this pathway is another potential pan-eukaryotic drug target (Overath and Engstler, 2004). These are just a few of those areas already under intense scrutiny. Other known drug targets for kinetoplastid metabolic processes include the pentose phosphate pathway, polyamine metabolism, protein degradation, purine salvage and pyrimidine biosynthesis (Barrett *et al.*, 2003). The unique organelle containing the mitochondrial DNA, the kinetoplast, also has the potential to be disrupted in several ways. The structural integrity could be disrupted using inhibitors of DNA topoisomerases (Das *et al.*, 2004), the biosynthesis of membrane components are targets using lipid synthesis inhibitors and the components of the RNA editing complex that are unique to these organisms, are also appropriate targets (Deng *et al.*, 2005).

It would seem that the very uniqueness of these organisms offers a gold mine of potential targets for pan-kinetoplastid drugs. However, realistically it would seem that the diverse and complex way the pathogens interact with the human host will limit the effectiveness of any such drugs. Both *L. major* and *T. cruzi* remain intracellular in the host, occupying different compartments of the cell meaning that an effective drug will have to cross into these cells to target the parasite. In African trypanosomiasis, many infections are not identified until the second stage when the parasites have reached the central nervous system, and therefore any drugs to treat this must also cross the blood–brain barrier. Ideally, to maximise the usefulness a drug should be available to be taken orally. One of the drawbacks of most of the frontline treatments for these diseases is that they require daily injections or infusions.

On a positive note, drugs do not necessarily have to be targeted to metabolic or structural processes present in the parasites, but absent or totally divergent in the host. Eflornithine, the best drug against second-stage trypanosomiasis in *T. gambiense* infections inhibits ornithine decarboxylase, which is present in humans and in trypanosomes. Its successful use relies on the fact that trypanosomes break down the compound at a much slower rate than mammalian cells and as such are much more susceptible to its toxic effects (Ghoda *et al.*, 1990).

Given the plethora of possible drug targets, how is the availability of the sequence data going to aid researchers and drug companies in the search for new drugs and effective vaccines? Despite exhaustive similarity searches to public databases, motif searches against domain databases and the combined efforts of hundreds of scientists in reviewing the genome data, approximately 50% of the genes still have no known function. There are still many metabolic processes, signalling pathways and structural proteins that have yet to be discovered. It is likely that a large proportion of these genes of unknown function will either code for unique kinetoplastid pathways or interact with the host and are therefore adapted specifically for parasite survival in the host and vector.

10   Beyond sequencing

The genome projects have provided the complete genetic complement for each of these human pathogens. As well as the advantage of allowing researches to identify complete pathways and systems it will also enable high-throughput whole genome-wide methodologies to provide answers to many of the practical issues which can be applied eventually to improve treatment regimes. Questions relating to drug resistance mechanisms, stage-specific expression, targets of available drugs, genes that are crucial for virulence, protective parasite antigens for use in vaccines, host immune evasion and functions of novel pathways can all be addressed. There are already a range of high-throughput projects utilising the genome data and adding to the knowledge derived *in silico* from the Tritryp genome sequencing and comparative analysis (*Figure 1*). A brief summary of some of the projects that have utilised data from the genome projects is given below.

10.1   *Protein structure and crystallisation*

Sequence data alone give limited possibilities when it comes to utilising the proteins for therapeutic or vaccination purposes. Determining the three-dimensional structure

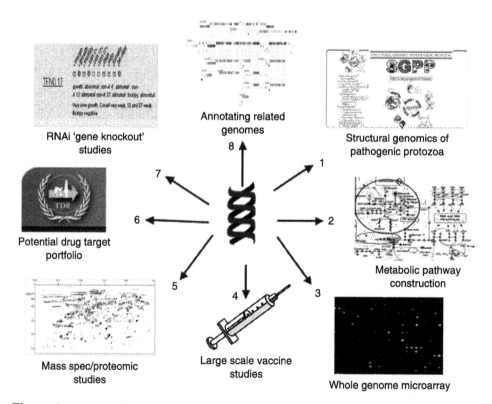

Figure 1. Projects utilising the genome data.

and crystallising the protein opens up a whole new dimension in research. The Structural Genomics of Pathogenic Protozoa (SGPP) project is a multicentre initiative that is attempting to systematically analyse the structure and solubility of proteins from each of the Tritryp genomes (www.sgpp.org). The project will attempt to quantify expression levels and use two-hybrid methods to work out the structures of soluble heteromultimers. Wherever possible the protein will also be crystallised.

10.2 Elucidating parasite pathways

As well as populating the KEGG metabolic pathway map with automatically predicted proteins derived from GO prediction program and Gene Ontology to EC number mapping as previously described, a suite of tools is being used to create a queriable pathway/genome database from the data generated from the genome projects. The Pathway Tools software (Karp et al., 2002) integrates information on the genes, proteins and metabolic and structural networks (http://bioinformatics.ai.sri.com/ptools/). The resulting database can be viewed in a graphical way, allowing users to visualise biochemical networks, and it can be extensively queried and mined for information.

10.3 Whole genome microarrays

These parasites undergo incredible morphological, structural and physiological changes during their life cycle. For example the environment in the sandfly is completely different to the extremes faced by the leishmania amastigote in the acidic parasitopherous vacuole in the mammalian host. The procyclic form that undergoes rapid cell division in the insect vector is structurally unable to survive attack from the human host's immune system. Extensive morphological and physical changes allow the metacyclic promastigote to survive this hostile environment. By comparing whole genome arrays of infective versus non-infective or virulent versus non-virulent parasites it is possible to identify those proteins that are differentially expressed in this transition. This methodology can also be applied to other aspects where proteins are differentially expressed. For example in comparing drug-resistant strains against drug-susceptible ones. There are a number of whole genome arrays in development for all these pathogens (Diehl et al., 2002; Holzer et al., 2006). Initial results suggest that very few genes are significantly differentially expressed in the life cycle stages. This may reflect the reliance on post-transcriptional protein regulation.

10.4 Two-dimensional gel/ mass spectrophotometry studies

Similarly to microarray analysis, an expression profile of the proteome can be derived from using two-dimensional (2D) gel electrophoresis. There are limitations in that membrane-bound and insoluble protein cannot be run and it is sometimes impossible to resolve individual proteins from dense spots. By combining mass spectrophotometry and data from the genome project it is possible to identify large numbers of proteins extracted from these gels resulting in a proteomic map. Composite 2D gels generated under different conditions resolved proteins from the L. major genome (Drummelsmith et al., 2003). Comparison of drug-resistant or virulent strains against the profile of wild-type organisms can help identify those proteins that are over expressed in these parasites. Methotrexate-resistant *L. major* parasites were found to over express the pteridine reductase PTRI gene, providing confirmation that this protein was involved in the mechanism of resistance (Drummelsmith *et al.*, 2003). It is known that most regulation of protein expression comes from post-translational modifications of proteins. Separation of sequences that are later identified as the same protein but with different molecular weights gives some insight into this process.

10.5 *Vaccine studies*

The comprehensive annotation of the genome data has given researchers a good source of prospective vaccine candidates. The long-term immunity in cured patients with leishmaniasis has encouraged researchers to persevere with the search for an effective vaccine against this disease. Pooled antigens from 100 leishmania proteins have been used in an attempt to scale up the testing of antigens to determine which proteins, either singularly or in combination, will elicit a protective cell-mediated immune response. Equally, this type of high-throughput strategy can identify those proteins that appear to exacerbate susceptibility to infection (Stober *et al.*, 2005). The advent of DNA vaccines and fusion of recombinant parasite antigens has shown some degree of success in murine and simian models of cutaneous infection and is undergoing initial trials in humans (Coler and Reed, 2005).

10.6 *Drug target portfolios*

With the publication of the genome sequences and consequent generation of 'postgenomic' data, a wealth of information has been created that needs to be linked to experimental data on therapeutic compounds that are already in the public domain. The Genomics and Discovery Research unit (GDR) of the UNICEF/UNDP/World Bank/WHO Special Programme for Research and Training in Tropical Diseases (TDR) has funded scientists to centralise all this information into a drug target portfolio that will be open source and accessible to anyone in the public domain (http://www.who.int/tdr/grants/grants/drug_parasitic.htm). The Tropical Biominer Project is another centralised database that is being designed for the purpose of marrying together information from the genome projects and drug discovery data with the aim of identifying known drugs for use against parasitic diseases (Artiguenave *et al.*, 2005).

10.7 *Gene knockout and knockdown*

Deleting the gene or inactivating the protein is an effective way to understand its function. Phenotypes that show a reduced or loss of capability to infect or survive in the host are important in identifying potential new targets for chemotherapy. Knocking out or interrupting both copies of a gene is an expensive and long procedure. One way to try and speed up the process is to use RNA interference (RNAi) in which the generation of double-stranded RNA molecules degrades the RNA of the specific protein *in vivo*, in effect acting as a temporary gene knockout. African trypanosomes were one of the first organisms in which RNAi was identified (Ngo *et al.*, 1998) but strangely it is one of the interesting differences between the three members of the Tritryps organisms. While it has been actively used against

African trypanosomes, it has consistently failed in both *T. cruzi* and the *Leishmania* spp. in which it has been tried. Comparison of the Tritryp genomes has revealed that *L. major* and *T. cruzi* are missing three genes thought to be part of the RNAi pathway present in *T. brucei*. The TrypanoFAN project (http://trypanofan.path.cam.ac.uk) is an attempt to catalogue the phenotypic data from a systematic knockdown of *T. brucei* genes. The project has started with the genes on chromosome 1 and RNAi mutants are measured by a series of structural, physiological and biochemical tests. The complexity of antigenic variation induced by VSG switching has also been analysed using RNAi (Aitcheson *et al.*, 2005).

10.8 *Annotating and assembling related genomes*

The high degree of manual finishing, additional sequencing in poorly covered or repetitive regions and the addition of restriction, genetic and optical maps have resulted in three high-quality essentially complete genomes. Along with the much smaller *Theileria annulata*, *L. major* is probably the most complete eukaryotic genome to date; it only lacks a single subtelomeric region of approximately 50 Kb to truly finish the sequence. The addition of manual annotation and curation on high-quality sequence has provided ideal templates on which to assemble and annotate the sequences of related genomes. This is especially useful in the kinetoplastid genomes given their high degree of conservation in genome architecture and gene order. Thus these high-quality reference genomes can be used to help assemble the contigs from the genome projects of other kinetoplastids and as such will allow for a lower level of sequence coverage and reduce the time taken to produce high-quality contiguous data. Despite considerable sequence divergence of orthologues, the availability of the finished genomes allows the additional information from synteny to be used to annotate genes, of particular use in these organisms, which are under represented in the public databases.

11 Future comparative sequencing projects

The kinetoplastid genome sequencing projects have now moved into comparing individual species or strains within these pathogenic groups. There are currently whole genome shotgun projects underway for additional members of all three genera in the Tritryps. Sequence data are already available from the genome sequencing initiatives for *Leishmania infantum*, *Leishmania brazilliensis*, *Trypanosoma congolense*, *Trypanosoma brucei gambiense* and *Trypanosoma vivax* and a second strain of *T. cruzi* has also been sequenced (El-Sayed *et al.*, 2005a). Sequencing additional species that cause disease in humans will help uncover which parasite genetic factors determine the tissue tropism, host immune response and drug response that typifies each specific disease. For example, in leishmaniasis certain species reside solely in the dermis and elicit a strong cell-mediated response in humans that leads to resolution of disease. Other species visceralise and mediate a generalised suppression of the hosts immune system. Host susceptibility factors in the mouse model have been shown to be different for each of the main pathogenic groups. Of the four main disease-causing groups in human leishmaniasis, a representative species of each has been chosen for sequencing, three of which are near completion. Another way to look for specific genes that enable successful survival of the pathogen in the human is to sequence members of the genus that are specific for a different host organism and are consequently unable to parasitise humans. Analysis of the genetic differences would highlight host specific virulence genes. Knockout and gene transfection studies could be used to identify the exact proteins from the limited list of species-specific genes, which could then be targeted for drug development. *Trypanosoma vivax* and *Trypanosoma congolense* are both pathogens of cattle and are unable to survive in humans. As sequencing costs decrease it will also become possible to follow the example of the sequencing of *Plasmodium* spp. and generate low-level sequence coverage of clinical isolates of an individual species that shows differential patterns of infection in humans (http://www.sanger.ac.uk/Projects/P_falciparum/). For example, some viscerotropic isolates of *Leishmania infantum* are known to cause cutaneous pathology while other isolates remain solely systemic (Sulahian *et al.*, 1997). These isolates can be differentiated by isoenzyme profiles and geographical location but are considered to be the same species.

Leishmaniasis, sleeping sickness and Chagas disease are three of the most important tropical infectious diseases in terms of human mortality and morbidity. They are considered a priority for research by both the World Health Organization and funding organisations and yet advances in treatment have lagged well behind other diseases. Development of drugs has been hindered by the economic challenges faced by drug companies needing to recoup development costs and the lack of understanding of the pathogens themselves. Publication of the genome sequences of a representative member from each of these pathogenic groups will allow the employment of genome-wide high-throughput post-genomic research to help maximise the investment available for drug development. Additionally, the high degree of conservation of core pathways as shown by comprehensive comparative genome analysis will provide hope for the development of broad-spectrum chemotherapeutic agents that can be managed within the limited resources many of the patients face. Deciphering the complex metabolic pathways will also help identify the targets that current drugs interact with and hopefully reveal some of the mechanisms that lead to drug resistance. It will also allow for *in silico* predictions to be made on the possibility that drugs developed for other diseases may be effective against trypanosomes. The development of a vaccine will provide the ultimate weapon against these diseases. There is no wholly effective vaccine against any parasitic disease that requires a strong cell-mediated response for disease resolution. However, with the observation of long-term immunity in patients cured of leishmaniasis, the advent of DNA vaccines that are better at stimulating the cell-mediated immunity than conventional vaccines and the complete complement of parasite genes it is hoped advances will be made in this area.

The Tritryp genome projects represent a new era in research of these diseases. It also represents a productive fusing of the work of sequencing centres and members of the research community. It is telling that despite the fact that the sequencing occurred in four centres, publication of the three genomes involved 237 authors from 46 institutions in 21 countries on 6 continents. The impact of the availability of these data can be reflected in the papers being cited 67 times in just 6 months since publication. The annotated sequence data for the Tritryp genomes and the additional kinetoplastid genomes currently being sequenced can be found in the GeneDB database (www.genedb.org).

References

Aitcheson, N., Talbot, S., *et al.* (2005) VSG switching in *Trypanosoma brucei*: antigenic variation analysed using RNAi in the absence of immune selection. *Mol Microbiol* **57(6)**: 1608–1622.

Andrade, L.O. and Andrews, N.W. (2005) The *Trypanosoma cruzi*-host-cell interplay: location, invasion, retention. *Nat Rev Microbiol* **3(10)**: 819–23.

Arce-Fonseca, M., Ballinas-Verdugo, M.A., *et al.* (2005) Autoantibodies to human heart conduction system in Chagas' disease. *Vector Borne Zoonotic Dis* **5(3)**: 233–6.

Artiguenave, F., Lins, A., *et al.* (2005) The Tropical Biominer Project: mining old sources for new drugs. *Omics* **9(2)**: 130–8.

Barrett, M.P., Burchmore, R.J., *et al.* (2003) The trypanosomiases. *Lancet* **362(9394)**: 1469–80.

Berriman, M., Ghedin, E., *et al.* (2005) The genome of the African trypanosome *Trypanosoma brucei*. *Science* **309(5733)**: 416–22.

Blackwell, J.M., Mohamed, H.S., *et al.* (2004) Genetics and visceral leishmaniasis in the Sudan: seeking a link. *Trends Parasitol* **20(6)**: 268–74.

Brun, R., Schumacher, R., *et al.* (2001) The phenomenon of treatment failures in Human African Trypanosomiasis. *Trop Med Int Health* **6(11)**: 906–14.

Burri, C. and Brun, R. (2003) *Human African Trypanosomiasis*. Saunders.

Coler, R.N. and Reed, S.G. (2005) Second-generation vaccines against leishmaniasis. *Trends Parasitol* **21(5)**: 244–9.

Croft, S.L., Barrett, M.P. *et al.* (2005) Chemotherapy of trypanosomiases and leishmaniasis. *Trends Parasitol* **21(11)**: 508–12.

Croft, S.L., Sundar, S. *et al.* (2006) Drug resistance in leishmaniasis. *Clin Microbiol Rev* **19(1)**: 111–26.

Cruz, A.K., Titus, R. *et al.* (1993) Plasticity in chromosome number and testing of essential genes in *Leishmania* by targeting. *Proc Natl Acad Sci USA* **90(4)**: 1599–603.

Das, A., Dasgupta, A. *et al.* (2004) Topoisomerases of kinetoplastid parasites as potential chemotherapeutic targets. *Trends Parasitol* **20(8)**: 381–7.

Dedet, J.P. and Pratlong, F. (2003) *Leishmaniasis*. Saunders.

Deng, J., Ernst, N.L. *et al.* (2005) Structural basis for UTP specificity of RNA editing TUTases from *Trypanosoma brucei*. *Embo J* **24(23)**: 4007–17.

Diehl, S., Diehl, F. *et al.* (2002) Analysis of stage-specific gene expression in the bloodstream and the procyclic form of *Trypanosoma brucei* using a genomic DNA-microarray. *Mol Biochem Parasitol* **123(2)**: 115–23.

Donelson, J.E. (2003) Antigenic variation and the African trypanosome genome. *Acta Trop* **85(3)**: 391–404.

Donn, R.P. and Ray, D.W. (2004) Macrophage migration inhibitory factor: molecular, cellular and genetic aspects of a key neuroendocrine molecule. *J Endocrinol* **182(1)**: 1–9.

Drummelsmith, J., Brochu, V. *et al.* (2003) Proteome mapping of the protozoan parasite *Leishmania* and application to the study of drug targets and resistance mechanisms. *Mol Cell Proteomics* **2(3)**: 146–55.

Eibl, H. and Unger, C. (1990) Hexadecylphosphocholine: a new and selective antitumor drug. *Cancer Treat Rev* **17(2–3)**: 233–42.

El-Sayed, N.M., Myler, P.J. *et al.* (2005a) The genome sequence of *Trypanosoma cruzi*, etiologic agent of Chagas disease. *Science* **309(5733):** 409–15.

El-Sayed, N.M., Myler, P.J. *et al.* (2005b) Comparative genomics of trypanosomatid parasitic protozoa. *Science* **309(5733):** 404–9.

Fairlamb, A.H., Blackburn, P. *et al.* (1985) Trypanothione: a novel bis(glu-tathionyl)spermidine cofactor for glutathione reductase in trypanosomatids. *Science* **227(4693):** 1485–7.

Ferguson, M.A., Brimacombe, J.S. *et al.* (1999) The GPI biosynthetic pathway as a therapeutic target for African sleeping sickness. *Biochim Biophys Acta* **1455(2–3):** 327–40.

Ghoda, L., Phillips, M.A. *et al.* (1990) Trypanosome ornithine decarboxylase is stable because it lacks sequences found in the carboxyl terminus of the mouse enzyme which target the latter for intracellular degradation. *J Biol Chem* **265(20):** 11823–6.

Girones, N., Cuervo, H. *et al.* (2005) *Trypanosoma cruzi*-induced molecular mimicry and Chagas' disease. *Curr Top Microbiol Immunol* **296:** 89–123.

Holzer, T.R., McMaster, W.R. *et al.* (2006) Expression profiling by whole-genome interspecies microarray hybridization reveals differential gene expression in pro-cyclic promastigotes, lesion-derived amastigotes, and axenic amastigotes in *Leishmania mexicana*. *Mol Biochem Parasitol* **146:** 198–218.

Ivens, A.C., Peacock, C.S. *et al.* (2005) The genome of the kinetoplastid parasite, *Leishmania major*. *Science* **309(5733):** 436–42.

Karp, P.D., Paley, S. *et al.* (2002) The Pathway Tools software. *Bioinformatics* **18** Suppl 1: S225–32.

Liang, X.H., Haritan, A. *et al.* (2003) Trans and cis splicing in trypanosomatids: mechanism, factors, and regulation. *Eukaryot Cell* **2(5):** 830–40.

Martin, D.M., Berriman, M. *et al.* (2004) GOtcha: a new method for prediction of protein function assessed by the annotation of seven genomes. *BMC Bioinformatics* **5:** 178.

Martinez-Calvillo, S., Nguyen, D. *et al.* (2004) Transcription initiation and termina-tion on *Leishmania major* chromosome 3. *Eukaryot Cell* **3(2):** 506–17.

Matsumoto, S.C., Labovsky, V. *et al.* (2006) Retinal dysfunction in patients with chronic Chagas' disease is associated to anti-*Trypanosoma cruzi* antibodies that cross-react with rhodopsin. *Faseb J.* **20(3):** 550–2.

Melville, S.E., Leech, V. *et al.* (1998) The molecular karyotype of the megabase chro-mosomes of *Trypanosoma brucei* and the assignment of chromosome markers. *Mol Biochem Parasitol* **94(2):** 155-73.

Michels, P.A., Hannaert, V. *et al.* (2000) Metabolic aspects of glycosomes in try-panosomatidae – new data and views. *Parasitol Today* **16(11):** 482–9.

Miles, M.A. (2003) *American Trypanosomiasis (Chagas Disease)*. Saunders.

Morita, Y.S., Paul, K.S. *et al.* (2000) Specialized fatty acid synthesis in African try-panosomes: myristate for GPI anchors. *Science* **288(5463):** 140–3.

Myler, P.J., Audleman, L. *et al.* (1999) *Leishmania major* Friedlin chromosome 1 has an unusual distribution of protein-coding genes. *Proc Natl Acad Sci USA* **96(6):** 2902–6.

Ngo, H., Tschudi, C. *et al.* (1998) Double-stranded RNA induces mRNA degrada-tion in *Trypanosoma brucei*. *Proc Natl Acad Sci USA* **95(25):** 14687–92.

Overath, P. and Engstler, M. (2004) Endocytosis, membrane recycling and sorting of GPI-anchored proteins: *Trypanosoma brucei* as a model system. *Mol Microbiol* 53(3): 735–44.

Paul, K.S., Bacchi, C.J. *et al.* (2004) Multiple triclosan targets in *Trypanosoma brucei*. *Eukaryot Cell* 3(4): 855–61.

Pepin, J. and Milord, F. (1991) African trypanosomiasis and drug-induced encephalopathy: risk factors and pathogenesis. *Trans R Soc Trop Med Hyg* 85(2): 222–4.

Simpson, A.G., Gill, E.E. *et al.* (2004) Early evolution within kinetoplastids (euglenozoa), and the late emergence of trypanosomatids. *Protist* 155(4): 407–22.

Solbach, W. and Laskay, T. (2000) The host response to *Leishmania* infection. *Adv Immunol* 74: 275–317.

Stober, C.B., Lange, U.G. *et al.* (2005) From genome to vaccines for leishmaniasis: screening 100 novel vaccine candidates against murine *Leishmania major* infection. *Vaccine.* 24(14): 2602–16.

Sukumaran, B. and Madhubala, R. (2004) Leishmaniasis: current status of vaccine development. *Curr Mol Med* 4(6): 667–79.

Sulahian, A., Garin, Y.J. *et al.* (1997) Experimental pathogenicity of viscerotropic and dermotropic isolates of *Leishmania infantum* from immunocompromised and immunocompetent patients in a murine model. *FEMS Immunol Med Microbiol* 17(3): 131–8.

Teixeira, M.J., Teixeira, C.R. *et al.* (2006) Chemokines in host-parasite interactions in leishmaniasis. *Trends Parasitol* 22(1): 32–40.

Turner, C.M., Melville, S.E. *et al.* (1997) A proposal for karyotype nomenclature in *Trypanosoma brucei*. *Parasitol Today* 13(1): 5–6.

Wang, C.C. (1995) Molecular mechanisms and therapeutic approaches to the treatment of African trypanosomiasis. *Annu Rev Pharmacol Toxicol* 35: 93–127.

Wincker, P., Ravel, C. *et al.* (1996) The *Leishmania* genome comprises 36 chromosomes conserved across widely divergent human pathogenic species. *Nucleic Acids Res* 24(9): 1688–94.

3

The relevance of host genes in malaria

Miguel Prudêncio, Cristina D. Rodrigues and Maria M. Mota

1 Introduction

Malaria is a devastating disease that affects extensive areas of Africa, Asia and South and Central America, causing up to 2.7 million deaths per year, mainly children under the age of five (Webster and Hill, 2003). The disease is caused by a protozoan parasite from the genus *Plasmodium* and transmitted through the bite of the female *Anopheles* mosquito. When a mosquito infected with *Plasmodium* bites a mammalian host, it probes for a blood source under the skin and, during this process, deposits saliva containing sporozoites. These sporozoites reach the circulatory system and are transported to the liver. Once there, they migrate through several hepatocytes by breaching their plasma membranes before infecting a final cell with the formation of a parasitophorous vacuole. After several days of development inside a hepatocyte, thousands of merozoites are released into the bloodstream where they invade red blood cells (RBCs), initiating the symptomatic erythrocytic stage of the disease (*Figure 1*).

Malaria infection depends upon the occurrence of interactions between the *Plasmodium* parasite and the host. Every stage of an infection by *Plasmodium* relies, to different extents, on the presence of host molecules that enable or facilitate its invasion, survival and multiplication. Therefore, host genes play a crucial role in determining the resistance or susceptibility to malaria and may constitute potential targets for preventive or therapeutic intervention. Analysis of the genetic basis of susceptibility to major infectious diseases is, arguably, the most complex area in the genetics of complex disease (Hill, 2001). In this chapter, we will examine the progress made towards identifying mammalian host molecules that play a role in the modulation of malaria infections.

2 The pre-erythrocytic stage: hepatocyte, liver and beyond

The hepatic stage of a *Plasmodium* infection constitutes an appealing target for the development of an intervention strategy since this would act before the onset of pathology, which only occurs during the blood stage of the parasite's life cycle. In fact, until now, the only demonstrably effective vaccine shown to confer a sterile and lasting protection both in mice (Nussenzweig *et al.*, 1967) and in humans (Clyde *et al.*, 1973;

Comparative Genomics and Proteomics in Drug Discovery, edited by John Parrington and Kevin Coward.
© 2007 Taylor and Francis Group.

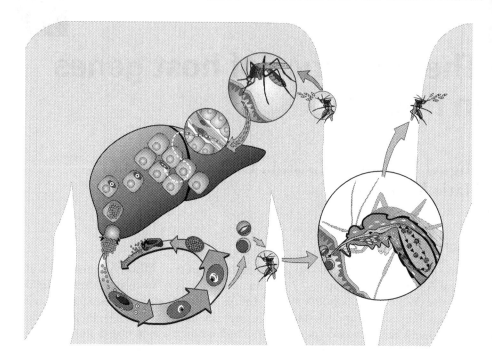

Figure 1. *The life cycle of* Plasmodium *in the mammalian and mosquito hosts.*

Rieckmann *et al.*, 1974; Herrington *et al.*, 1991) was the inoculation of γ-irradiation-attenuated sporozoites, that are able to invade but not fully mature inside the hepato-cyte (see Carvalho *et al.*, 2002; Gruner *et al.*, 2003; Bodescot *et al.*, 2004; Todryk and Walther, 2005; Waters *et al.*, 2005).

Despite being symptomatically silent, the liver stage of a malaria infection is immuno-logically very complex. Unlike RBCs, liver cells are able to promote cell-mediated immune response mechanisms through expression of class I Major Histocompatibility Complex (MHC) proteins. Class I MHC proteins present antigens to cytotoxic T lymphocytes (CTLs) (Lowell, 1997), which are known to play an important role in the generation of a protective immune response in many microbial infections (Esser *et al.*, 2003). The acti-vation of T cells by antigen-presenting cells (APCs) is required to initiate specific immune responses. Different APCs have been shown to be important in this process, including dendritic cells (DC) (Bruna-Romero and Rodriguez, 2001; Jung *et al.*, 2002; Leiriao *et al.*, 2005) and Kupffer cells (Steers *et al.*, 2005).

Human leucocyte antigens (HLAs) are encoded by genes of the MHC, which are known to be among the most polymorphic of all human genes (reviewed in Williams, 2001). Although most of the human MHC *loci* are relatively stable, the *HLA-B locus* has been shown to undergo rapid changes, especially in isolated populations (McAdam *et al.*, 1994). *HLA-B* encodes an MHC class I heavy chain that is part of the *HLA-B* antigen-presentation complex (Kwiatkowski, 2005). On the other hand, the *HLA-DR* antigen-presenting complex includes an HLA class II β chain, which is encoded by *HLA-DRB1*. HLA-DR is found in B lymphocytes, DC and macrophages where it plays an essential role in the production of antibodies (Kwiatkowski, 2005). Both the *HLA-B53* allele and the *DRB1*1302-DQB1*0501* haplotype were shown to be associated with protection against severe malaria (SM) in The Gambia (Hill *et al.*, 1991) and the latter was also found to be associated with protection from malaria anaemia and malarial reinfections in Gabonese children (May *et al.*, 2001).

The analysis of the peptides from a vast range of malaria antigens that bind to HLA-B53-restricted CTLs in malaria-immune Africans led to the identification of a single conserved peptide from liver-stage-specific antigen-1 (LSA-1) (Hill *et al.*, 1992), making LSA-1 an interesting malaria vaccine candidate (Migot-Nabias *et al.*, 2001). Of the four most prevalent allelic variants of the protein recognized by HLA-B53, only two are indeed epitopes, binding this protein *in vitro* (Gilbert *et al.*, 1998). Moreover, these results suggest that cohabiting parasite strains, each of which being an individually effective target for CTLs, may have the ability to use altered peptide ligand (APL) antagonism mechanisms to suppress the CTL response to the other strain, thereby increasing each other's chances of survival. This observation has obvious implications in terms of vaccine development, since it suggests that including all allelic peptide variants in a prospective vaccine might be counterproductive because one given variant may antagonize immunity to other variants.

During *Plasmodium* sporozoite development inside hepatocytes there is an amazing multiplication, with each parasite giving rise to 10,000–30,000 merozoites in 2–7 days (depending on *Plasmodium* spp.). Moreover, there is a high level of specificity of *Plasmodium* sporozoite development, which only occurs in certain types of cells. This strongly suggests an important role for the host cell in supporting the full development of the parasite. However, not much is known about *Plasmodium* requirements, strategies developed to survive and be successful, or how much the host cell contributes to this. An intriguing characteristic of *Plasmodium* sporozoites is their ability to migrate through hepatocytes prior to invading a final one with the formation of a vacuole for further development. Sporozoites breach the plasma membrane of the cell, traverse through its cytosol and leave by wounding the membrane (Mota *et al.*, 2001). This unusual process is frequently observed *in vitro*, where sporozoites traverse mammalian cells at a speed of approximately one cell per minute. Migration through host hepatocytes is also observed *in vivo* in the liver of mice infected with *Plasmodium* sporozoites (Mota *et al.*, 2001; Frevert *et al.*, 2005). Wounding of host hepatocytes induces an alteration in traversed cells which includes the secretion of host cell factor(s), which render(s) neighbouring hepatocytes susceptible to infection. One such factor is hepatocyte growth factor (HGF), which, by activating its receptor MET, seems to be required for the early development of parasites within host cells (Carrolo *et al.*, 2003). These results, however, appear to be contradicted by evidence provided by *spect*-deficient sporozoites, which do not migrate through cells *in vitro* (Ishino *et al.*, 2004). It would be expected that *spect*-deficient sporozoites would not induce host cells to produce HGF and, therefore, infection would be inhibited. However, these sporozoites efficiently infect host cells *in vitro* (Ishino *et al.*, 2004). This apparent discrepancy may be due to particular characteristics of the *in vitro* cell system used for infection (Mota and Rodriguez, 2004), or to alternative ways used by the parasite to fully develop inside host cells.

3 The erythrocytic stage and disease

The erythrocytic stage of *Plasmodium*'s life cycle corresponds to the symptomatic phase of a malaria infection. During this phase, *Plasmodium* merozoites invade RBCs and degrade haemoglobin (Hb), releasing heme that is converted into haemozoin (malarial pigment). The invading merozoites multiply in the RBCs and, upon ruptur-ing the erythrocytic membrane, are eventually released into the blood where they target new RBCs. The interaction between *Plasmodium* and the RBCs occurs in two stages: first, the identification and binding of RBC surface molecules that will enable invasion; subsequently, the intracellular interaction with Hb and the multiplication of the parasite.

3.1 *Erythrocyte invasion* – P. vivax *versus* P. falciparum

The human malaria parasite, *P. vivax*, and the related monkey malaria, *P. knowlesi*, use the Duffy blood group antigen as a receptor to invade human RBCs (Miller *et al.*, 1976). The Duffy antigen, encoded by the *FY* gene, is a chemokine receptor on the surface of the RBCs. It belongs to the superfamily of G-protein coupled receptors (GCRs) (Neote *et al.*, 1994) and is also termed DARC (for Duffy antigen receptor for chemokines). The Duffy blood group locus is polymorphic and has three main alleles designated *FY*A*, *FY*B* and *FY*O*. The *FY*A* allele is very frequent in Asia and the Pacific, whereas in Europe and the Americas the *FY*A* and *FY*B* alleles are at inter-mediate frequencies. The *FY*O* allele is at or near fixation in most sub-Saharan African populations, but is very rare outside Africa (Hamblin and Di Rienzo, 2000). *FY*O* arises from a mutation at position -46 in the promoter of the *FY*B* allele, leading to a Fy(a-b-) phenotype in which RBCs lack both Fy^a and Fy^b antigens (Tournamille *et al.*, 1995). Fy(a-b-) RBCs resist invasion *in vitro* by *P. knowlesi* parasites and indi-viduals homozygous for the *FY*O* allele are completely resistant to *P. vivax* malaria (Miller *et al.*, 1976). The correlation between the Duffy-negative serological pheno-type and resistance to *P. vivax* malaria is now clear. *P. vivax* relies on a single path-way to invade RBCs. The lack of redundancy in *P. vivax* invasion pathways may explain the near absence of this parasite from West Africa, where almost 95% of the population have the Duffy-negative phenotype and are resistant to *P. vivax* malaria (Chitnis and Miller, 1994). The extreme degree of between-population differentiation of allele frequency of the Duffy blood group gene shows evidence of human directional selection by *P. vivax* (Hamblin *et al.*, 2002). Furthermore, it has important implica-tions in drug or vaccine design (Yazdani *et al.*, 2004).

Unfortunately, things are a lot more complicated when it comes to the much more deadly *P. falciparum* malaria parasite.

Contrary to *P. vivax*, *P. falciparum* displays the ability to invade RBCs following multiple, alternative pathways, with significant redundancy. Research into the identifi-cation of RBC receptors involved in merozoite invasion has made use of RBCs that lack specific surface molecules or enzymes that modify protein and carbohydrate domains on those molecules. The main enzymes used have been neuraminidase (which cleaves sialic acid groups from surface glycoproteins and glycolipids) and trypsin (which cleaves the peptide backbone of a number of surface proteins) (Baum *et al.*, 2003). These studies have revealed several surface receptors that are involved in invasion of the RBC by *P. falciparum* merozoites. Three neuraminidase-sensitive molecules that have received particular attention are glycophorin A (GYPA) (Pasvol *et al.*, 1982), glycophorin B (GYPB) (Dolan *et al.*, 1994), and glycophorins C and D (GYPC/D) (Mayer *et al.*, 2001; Maier *et al.*, 2003). GYPC and GYPD are encoded by the same gene, but use alternative start codons. Deletion of exon 3 in the *GYPC/D* gene changes the serologic phenotype of the Gerbich (Ge) blood group system, resulting in Ge-negativity. Ge-negative RBCs exhibit a shortened GYPC and lack GYPD (Mayer et al., 2001). This is of particular relevance if we consider that the Ge-negative phenotype is found at high allele frequencies in some regions of Papua New Guinea, which coincide with regions of malaria hyperendemicity. This strongly suggests that selection of Ge-negativity in these populations confers at least partial protection against P. falciparum malaria. It should be noted that this is in contrast to a study in which no correlation between the prevalence of P. falciparum infection and Ge-negativity in the Wosera region of Papua New Guinea was found (Patel et al., 2001). More recently, the existence of yet another sialic acid-dependent receptor, termed 'Receptor Y', has been demonstrated (Rayner et al., 2001). It would seem apparent that P. falciparum shows a near-exclusive preference for sialic acid-dependent (i.e. neuraminidase-sensitive) glycophorins for RBC invasion. This, however, is now known not to be the case. Several reports have shown that alternative, sialic acid-independent, invasion pathways are commonly used by P. falciparum (Mitchell et al., 1986; Hadley et al., 1987; Dolan et al., 1994). Moreover, P. falciparum seems capable not only of sialic acid-independent RBC invasion, but also of switching between sialic acid-dependent and sialic acid-independent pathways (Dolan et al., 1990; Reed et al., 2000; Duraisingh et al., 2003). Very recently, the molecular mechanism for this switching process was elucidated (Stubbs et al., 2005).

As we have seen, the surface of the RBC presents various types of sialic acid-dependent and sialic acid-independent receptors that can mediate P. falciparum invasion with at least a certain degree of redundancy. Furthermore, it is clear that the relative importance of each receptor is strain-dependent (Rayner et al., 2001). These observations have important implications for vaccine development. While the existence of a single RBC invasion pathway for P. vivax yields good reasons to hope for a successful vaccine, this is clearly not the case for P. falciparum.

3.2 Intraerythrocytic stage – globin and non-globin genes

Following invasion of RBCs, Plasmodium parasites develop and multiply, leading to the appearance of the symptoms of malaria infection. Almost 90% of the intraerythrocytic space is taken up by Hb. Therefore, it is to be expected that Plasmodium will interact closely with this molecule and be influenced by its overwhelming presence. Thus, it is perfectly conceivable that Hb alterations will affect the development of the parasite, as well as the parasitized RBC itself.

Normal Hbs are tetrameric proteins composed of two pairs of unlike globin chains. After birth, the vast majority of Hb is composed of two α-globin and two β-globin chains (reviewed in Weatherall and Clegg, 2001; Richer and Chudley, 2005). The molecular pathology of most of the haemoglobinopathies is well defined (Weatherall and Clegg, 2001). The Hb disorders resulting from mutations in the α- or β-globin gene clusters are the most common single-gene disorders in humans (Weatherall, 2001). Inherited haemoglobinopathies can be divided into two main groups: structural Hb variants, mostly resulting from single amino acid substitutions in the α- or β-chains, and thalassaemias, arising due to the ineffective synthesis of the α- and/or β-chains (reviewed in Weatherall and Clegg, 2001). These two classes of Hb disorders constitute one of the most striking illustrations of why malaria is considered the strongest known selective pressure in the recent history of the human genome (Kwiatkowski, 2005).

Structural Hb variants. Of the more than 700 structural Hb variants identified, only those coding for 'haemoglobin S' (or 'sickle haemoglobin', HbS), 'haemoglobin C' (HbC) and 'haemoglobin E' (HbE) have reached polymorphic frequencies. Each of the alleles *HbS*, *HbC* and *HbE* results from a single point mutation in the *HBB* gene. Although often lethal in the homozygous state, the unusually high prevalence of these alleles in areas of malaria endemicity has long been attributed to a selective pressure exerted by *Plasmodium* on the human host genome (Min-Oo and Gros, 2005).

In HbS, the glutamate at position 6 of the β-globin chain is replaced by a valine residue. The resulting protein contains 'sickle' β-globin (β^S-globin) chains and tends to polymerize at low oxygen concentrations, causing the RBC to acquire a sickle-like shape (Brittenham *et al.*, 1985). This results in a condition known as 'sickle cell' anaemia, an autosomal recessive genetic disorder characterized by chronic anaemia and periodic vaso-occlusive crises (Shiu *et al.*, 2000). Although the homozygous state (HbSS) is often lethal, the heterozygous state (HbAS) is referred to as sickle cell trait and is usually clinically silent. The HbAS state has been shown to confer significant protection to malaria (reviewed in Kwiatkowski, 2005; see also Aidoo *et al.*, 2002; Williams *et al.*, 2005a). The exact mechanisms through which HbAS protects against malaria are unclear. A few possible explanations have, however, been put forward, including the enhanced sickling of the infected RBCs (iRBCs) (Luzzatto *et al.*, 1970), the suppression of parasite growth in individuals with the sickling disorders (Pasvol *et al.*, 1978) and increased spleen clearance (Shear *et al.*, 1993), enhanced phagocytosis (Ayi *et al.*, 2004) and enhanced acquisition of natural immunity to malaria, possibly due to the accelerated acquisition of antibodies against altered host antigens expressed on the surface of the infected RBCs (iRBCs) and/or against parasite-derived proteins (Williams *et al.*, 2005b).

In HbC, the glutamate at position 6 of the β-globin chain is replaced by a lysine residue. The resulting condition is considerably less serious than sickle cell anaemia. Even in the homozygous state (HbCC), only occasional pathologic developments are observed whereas heterozygotes (HbAC) are asymptomatic (Agarwal *et al.*, 2000; Kwiatkowski, 2005). This Hb variant has been implicated in protection against malaria in both the HbAC and the HbCC states (reviewed in Kwiatkowski, 2005; Min-Oo and Gros, 2005), although contrasting results have been reported concerning the heterozygous state (discussed in Modiano *et al.*, 2001). Three recent reports propose complementary explanations for the protective effect of HbC against malaria, all implying modifications that occur at the host cell surface (Tokumasu *et al.*, 2005; Arie *et al.*, 2005; Fairhurst *et al.*, 2005).

HbE results from a glutamate → lysine mutation at position 27 of the β-globin chain. It is the most common structural variant of Hb and is innocuous both in its heterozygous (HbEA) and homozygous (HbEE) states (Weatherall and Clegg, 2001). There is no unequivocal proof that HbE protects against malaria but it has been suggested that an alteration in the RBC membrane in HbEA cells renders the majority of the RBCs population relatively resistant to invasion by *P. falciparum* (Chotivanich *et al.*, 2002).

Thalassaemias. The β- and α-thalassaemias are disorders of globin chain synthesis that appear as a consequence of deletions or point mutations in the non-coding portion of the β- and α- globin genes, respectively (reviewed in Weatherall, 2001; Richer and Chudley, 2005). Because the synthesis of the β-globin chain is determined by two alleles of the *HBB* gene, whereas that of α-globin is encoded by the four alleles on the equivalent *HBA1* and *HBA2* genes, there is a wide range of possible thalassaemic genetic variants, with different clinical manifestations. In general, homozygous thalassaemia is severe or even fatal, whereas the heterozygous state is clinically benign. However, when either the *HBA1* or the *HBA2* gene, but not both, is defective, a condition termed α-thalassaemia occurs, for which homozygous individuals show only mild anaemia (Kwiatkowski, 2005). Evidence for protection against malaria by thalassaemia has been shown in different reports (reviewed in Min-Oo and Gros, 2005; Williams *et al.*, 2005a). The nature of protection against malaria by thalassaemias is unclear. Again, several suggestions have been mentioned including impaired parasite growth and increased susceptibility to phagocytosis (Yuthavong *et al.*, 1988, 1990), altered expression of parasite-induced surface neoantigens in β- and α-thalassaemia allowing greater binding of specific antibody to iRBCs and their subsequent clearance (Luzzi *et al.*, 1991a, 1991b) and enhanced phagocytosis of ring-stage iRBCs of β- (but not α-) thalassaemia individuals (Ayi *et al.*, 2004).

In addition to Hb, other RBC molecules also seem to play an important role in the parasite's development and in the host's susceptibility to malaria.

G6PD deficiency. The haeme liberated in the process of intra-erythrocytic Hb degradation by *Plasmodium* is polymerized to haemozoin or broken down non-enzymatically. Non-enzymatic cleavage of haeme liberates iron and generates hydrogen peroxide, a potential source of oxidative stress (Schwarzer *et al.*, 2003). Glucose-6-phosphate dehydrogenase (G6PD) catalyses the oxidation of glucose-6-phosphate to 6-phosphogluconate, while concomitantly reducing nicotinamide adenine dinucleotide phosphate ($NADP^+$ to NADPH). G6PD is the only erythrocytic enzyme that produces NADPH, a compound that is crucial for the defence of the RBCs against oxidative stress (Beutler, 1996). G6PD deficiency is the most common human enzymopathy known, affecting over 400 million people (Beutler, 1990). There are numerous variants of the *G6PD* gene and only those that significantly hamper enzyme activity lead to haemolytic anaemia, a phenotype that is exacerbated under oxidative stress conditions (Kwiatkowski, 2005). Evidence for the geographical correlation between G6PD deficiency and protection against malaria comes from various population-based studies and suggests evolutionary selection by the latter (reviewed in Kwiatkowski, 2005; Min-Oo and Gros, 2005). Although reduced parasite replication in G6PD-deficient RBCs was initially proposed as the mechanism of protection (Luzzatto *et al.*, 1969), the invasion and maturation of the parasite seems not to be significantly different between normal and G6PD-deficient RBCs (Cappadoro *et al.*, 1998). Instead, ring-stage iRBCs present on their surface a higher density of phagocytic removal markers than normal iRBCs and are, therefore, phagocytozed more efficiently. This suggestion has been extended to explain the protective nature of other erythrocytic defects against malaria (Ayi *et al.*, 2004), as previously mentioned.

Band 3 protein. Band 3 protein, encoded by *AE1*, is a ~100 kDa membrane protein that acts as an anion permeation channel, exchanging intracellular bicarbonate for chloride through the RBC lipid bilayer (Rothstein *et al.*, 1976; Alper *et al.*, 2002).

A 27-base-pair deletion in the *AE1* gene (known as Band3Δ27), has been shown to cause a condition known as South-East Asian ovalocytosis (SAO), characterized by slightly oval or elliptical shaped RBCs, defective anion transport activity, increased acidosis and haemolytic anaemia (Schofield *et al.*, 1992). Whereas the homozygous state is lethal, population-based studies have shown that the heterozygous state confers protection against cerebral malaria (CM), a condition characterized by progressing coma, unconsciousness, multiple convulsions and, often, death (Rasti *et al.*, 2004; Genton *et al.*, 1995; Allen *et al.*, 1999). Again, the mechanism of protection is unknown but several hypothesis have arisen, such as membrane alterations in SAO RBCs affecting parasite invasion (Kidson *et al.*, 1981) and involvement of Band 3 protein in iRBC cytoadherence, a process described below (Winograd and Sherman, 1989; Winograd *et al.*, 2005; Shimizu *et al.*, 2005).

PK deficiency. RBCs do not have mitochondria and are, therefore, dependent on glycolysis for energy. Pyruvate kinase (PK) is a key enzyme in glucose metabolism, catalysing the conversion of phosphoenolpyruvate into pyruvate, with simultaneous production of ATP. Thus, PK-deficient RBCs have impaired glycolysis and difficulty in maintaining normal levels of ATP and NAD. The degree of severity of PK-deficiency depends on the mutation(s) in one or both genes coding for PK in the human genome (Min-Oo and Gros, 2005). No formal evidence exists of a protective effect of PK deficiency against malaria in humans (Min-Oo and Gros, 2005). However, such a correlation is likely to exist, as suggested by recent studies carried out in mouse models (Min-Oo *et al.*, 2003, 2004).

3.3 *Cytoadherence and sequestration – selective pressures and therapeutic value*

P. falciparum is the most virulent of all four *Plasmodium* parasites that infect humans. Two features are strongly suggested to contribute decisively to the outstanding pathogenicity of *P. falciparum*: its remarkable potential to multiply to high parasite burdens and its unique ability to cause iRBCs to adhere to the linings of small blood vessels. The latter process is termed cytoadherence and the ensuing sequestration of iRBCs is frequently suggested to be a key feature in the pathogenesis of SM (see Ho and White, 1999; Kyes *et al.*, 2001; Miller *et al.*, 2002 for reviews).

The most widely suggested justification for sequestration in *P. falciparum* malaria is that adhesion of iRBCs to the endothelium allows the parasite to escape peripheral circulation and be cleared by the spleen. Another *Plasmodium* survival advantage is that sequestration in the deep tissue microvasculature provides the parasites with a microaerophilic venous environment that promotes maturation and faster asexual replication (Cranston *et al.*, 1984). Alternative, but not necessarily exclusive, explanations include sheltering of the iRBCs against destruction by the immune system of the host, enhanced survival of the gametocyte and immunomodulation by inhibition of the maturation and activation of DC (reviewed in Sherman *et al.*, 2003).

It has been suggested that parasite accumulation in specific organs is an important factor in malaria pathogenesis. Indeed, post-mortem examinations of people who have died from *P. falciparum* malaria show sequestered iRBCs within the small vessels of several tissues. The best-documented situation regards the sequestration of iRBCs in the endothelium of small vessels of the brain, which can lead to CM

(Kyes et al., 2001; Rasti et al., 2004). The proportion of iRBCs was higher in P. falciparum malaria patients dying with CM than in those that showed no CM symptoms (MacPherson et al., 1985; Silamut et al., 1999; Taylor et al., 2004). The most prevalent hypothesis to explain the clinical manifestations of CM seems to be that sequestration obstructs blood flow, and, consequently, decreases the levels of oxygen and nutrients, whilst increasing the accumulation of waste products in the brain (Miller et al., 1994). A complementary hypothesis is that an inflammatory response to the infection activates leukocytes and promotes their adhesion to receptors in the endothelium (Sun et al., 2003). Another organ where sequestration seems to play a particularly important pathogenic role is the placenta. Maternal or placental malaria (PM) in P. falciparum-infected pregnant women is associated with disease and death of both mother and child. The characteristic feature of infection during pregnancy is the selective accumulation of iRBCs in the intervillous blood spaces of the placenta, leading to hypoxia, inflammatory reactions and intervillositis (reviewed in Beeson et al., 2001; Andrews and Lanzer, 2002; Duffy and Fried, 2003).

Besides the cytoadherence of iRBCs to endothelial cells (sequestration), other distinctive patterns of iRBCs adherence can be distinguished at the cellular level. These involve different types of interactions between iRBCs and other host cells, such as RBCs (rosetting) (Udomsangpetch et al., 1989; 1992), platelets ('platelet-mediated clumping', formerly known as autoagglutination) (Roberts et al., 1992; Pain et al., 2001a) or DC (Urban et al., 1999, 2001). All these different types of interactions have been suggested to have implications in the course of infection as well as in disease severity (Udomsangpetch et al., 1989; Urban et al., 1999; Pain et al., 2001a).

Over the last few years, significant progress has been made towards the identification of both parasite and host molecules that participate in these interactions. An array of host molecules have been identified that mediate cytoadherence of the iRBCs to various host cells. Among the most important ones identified so far are CD36 (Barnwell et al., 1985), thrombospondin (TSP) (Roberts et al., 1985), intercellular adhesion molecule-1 (ICAM-1) (Berendt et al., 1989), vascular cell adhesion molecule-1 (VCAM-1) (Ockenhouse et al., 1992), E-selectin (endothelial leukocyte adhesion molecule 1, ELAM-1) (Ockenhouse et al., 1992), chondroitin sulphate A (CSA) (Rogerson et al., 1995), platelet/endothelial cell adhesion molecule-1 (PECAM-1/CD31) (Treutiger et al., 1997), complement receptor 1 (CR1) (Rowe et al., 1997), hialuronic acid (HA) (Beeson et al., 2002) and heparan sulphate (HS) (Vogt et al., 2003). Nevertheless, the exact role of each of these molecules in pathogenesis remains largely unclear. Whilst some host ligands appear to be nearly ubiquitous, others seem to be organ- or cell-specific. This specificity has implications in terms of disease severity and adhesion mechanisms. Moreover, sequestration in an organ is likely to involve multiple receptors, and different combinations of specific receptors for adhesion may determine the site at which parasites adhere and accumulate (Beeson and Brown, 2002). Furthermore, receptors can act synergistically in mediating the adhesion of iRBCs (McCormick et al., 1997; Yipp et al., 2000; Heddini et al., 2001b) and in vivo sequestration has been shown to involve a multi-step adhesive cascade of events where iRBCs initially roll along the endothelial surface and are subsequently arrested (Ho et al., 2000).

CD36. CD36 (encoded by CD36), also known as GP88 or platelet glycoprotein IV, is an 88 kDa cell surface class B scavenger receptor and is expressed in endothelial cells, monocytes, platelets and erythroblasts (Barnwell *et al.*, 1989; Ho and White, 1999). CD36 has been implicated in the interaction with a variety of natural ligands (reviewed in Ho and White, 1999; Serghides *et al.*, 2003) and it was identified *in vitro* as a receptor for iRBCs (Barnwell *et al.*, 1985, 1989). Since then, evidence for the involvement of CD36 in malarial cytoadherence became abundant and unequivocal, although its implications in disease severity are not always clear-cut. Whilst some reports describe a contribution of CD36 to malaria severity (Udomsangpetch *et al.*, 1992; Pain *et al.*, 2001a; Urban *et al.*, 2001; Prudhomme *et al.*, 1996), others suggest that it might be advantageous for host survival (Rogerson and Beeson, 1999; Serghides *et al.*, 2003; McGilvray *et al.*, 2000; Traore *et al.*, 2000). Several *CD36* polymorphisms have been identified. In Africa, the most common of these substitutions is a T1264G stop mutation in exon 10 (Aitman *et al.*, 2000). Studies of the effects of this polymorphism in disease severity have again yielded contradictory results and do not provide definitive answers regarding a possible selective pressure by malaria in endemic areas (Aitman *et al.*, 2000; Pain *et al.*, 2001b; Omi *et al.*, 2003).

The role of CD36 in the pathogenicity of malaria is still ambiguous. This has important implications in terms of the development of therapies or vaccines that target the interaction between PfEMP-1 and the CD36 receptor. The *in vitro*-based assumption that adherence to CD36 contributes to disease severity, and results in negative clinical outcomes, triggered research aimed towards interfering with this interaction that did not always yield agreeing results (Barnwell *et al.*, 1985; Baruch *et al.*, 1997; Cooke *et al.*, 1998; Yipp *et al.*, 2003). One way to try and circumvent this problem is by making use of appropriate animal, ideally rodent, models. The sequences of human and rat CD36 differ in a single amino acid (His242 in humans is Tyr242 in rat) and both are able to bind *P. falciparum*-infected RBCs (Serghides *et al.*, 1998). It has been shown that RBCs infected with the murine parasite *P. chabaudi chabaudi* AS adhere *in vitro* to purified CD36 and are sequestered from circulation in an organ-specific way *in vivo* (Mota *et al.*, 2000). Recently, a novel system that enables real-time *in vivo* imaging luciferase-expressing rodent malaria parasite *P. berghei* was used to monitor sequestration (Franke-Fayard *et al.*, 2005). Using CD36$^{-/-}$ mice, it was shown that nearly all detectable iRBC sequestration depends on CD36 and that murine CM pathology still develops in the absence of this receptor, implying that CD36-mediated sequestration in nonerythroid organs does not constitute the molecular basis of rodent CM. Despite the obvious advantages of using animal models to study sequestration, caution should be employed when extrapolating results to the *P. falciparum*/human situation.

ICAM-1. Intercellular adhesion molecule-1 (ICAM-1), encoded by *ICAM-1*, is a surface glycoprotein member of the immunoglobulin superfamily. ICAM-1, also known as CD54, plays a central role in immune response generation by functioning as an endothelial and immune-cell ligand for integrin-expressing leukocytes (reviewed in Ho and White, 1999; Kwiatkowski, 2005). ICAM-1 is widely distributed in endothelial cells including, unlike CD36, those of the brain microvascular system (Adams *et al.*, 2000). Also unlike CD36, the expression of *ICAM-1* can be up-regulated by a number of factors, most notably pro-inflammatory cytokines such as tumour necrosis factor-α (TNF-α), interleukin-1 (IL-1) and interferon-γ (IFN-γ) (reviewed in Dietrich, 2002), which seem to correlate with disease severity.

P. falciparum iRBCs have been shown to bind with different affinities to ICAM-1 *in vitro* (Berendt *et al.*, 1989). Unequivocal *in vivo* evidence of the involvement of ICAM-1 in sequestration, mainly in the brain, has been reported in both humans (Turner *et al.*, 1994) and mice models (Willimann *et al.*, 1995; Kaul *et al.*, 1998). The evidence gathered in these studies has led to the notion that ICAM-1 may play an important role in CM by either sequestering iRBCs or adhering to activated leukocytes. Furthermore, this receptor has also been shown to contribute to the cytoadherence of iRBCs within the intervillous spaces of the *P. falciparum*-infected placenta, supporting a possible role of ICAM-1 in PM (Sugiyama *et al.*, 2001), as previously suggested (Sartelet *et al.*, 2000). Moreover, ICAM-1 can synergize with CD36 to promote iRBC adhesion under flow conditions (McCormick *et al.*, 1997, Ho *et al.*, 2000; Yipp *et al.*, 2000). For these reasons, it is generally accepted that ICAM-1 plays a role in iRBC sequestration and in the severity of disease.

A high-frequency coding polymorphism in the *ICAM-1* gene of individuals from Kilifi (Kenya), an area of high malaria endemicity, has been identified (Fernandez-Reyes *et al.*, 1997). The mutant protein, termed ICAM-1[Kilifi], contains a lysine → methionine replacement at position 29 (Fernandez-Reyes *et al.*, 1997). Intuitively, one would be led to think that the prevalence of this polymorphism in malaria-endemic regions would be associated with protection against severe forms of disease. However, studies attempting to correlate this polymorphism with malaria severity, yielded surprising and contradictory results (Fernandez-Reyes *et al.*, 1997; Kun *et al.*, 1999; Bellamy *et al.*, 1998; Ohashi *et al.*, 2001; Amodu *et al.*, 2005). The obvious difficulties that arise when attempting to reconcile those results constitute an illustration of the extreme complexity of malaria pathogenesis and suggest that different biological selective forces might simultaneously be at play in the studied populations (Craig *et al.*, 2000). In a very recent report, an ICAM-1 exon 6 polymorphism (lysine → glutamate replacement at position 469 in the protein) was seen to positively correlate with an increased risk of SM (Amodu *et al.*, 2005).

PECAM-1/CD31. Platelet/endothelial cell adhesion molecule-1 (PECAM-1, also known as CD31), encoded by *PECAM-1*, is a highly glycosylated transmembrane glycoprotein of the immunoglobulin superfamily (Newman *et al.*, 1990). It is expressed on endothelial cells and platelets, as well as on granulocytes, monocytes, neutrophils and naïve T lymphocytes (Newman *et al.*, 1990; Mannel and Grau, 1997).

The role of PECAM-1 as an endothelial receptor for adherence of *P. falciparum*-infected RBCs was demonstrated *in vitro* by Treutiger *et al.* (1997). These authors further demonstrated that binding could be blocked by monoclonal antibodies (mAbs) specific for the N-terminus of PECAM-1 whereas it could be increased by IFN-γ, a proinflammatory cytokine associated with the development of CM. In addition to serving as an endothelial receptor, the fact that PECAM-1 is expressed on platelets raises the possibility that it may also be involved in platelet adhesion of iRBCs, although no direct evidence exists for this. Platelets have been proposed as important effectors of neurovascular injury in CM, both in mice and in humans (reviewed in Mannel and Grau, 1997). Furthermore, the PECAM-1 receptor is quite commonly recognized by wild *P. falciparum* isolates (Heddini *et al.*, 2001a). In this particular study, over 50% of the fresh patient-isolates tested recognized PECAM-1, a number that is not matched by any other endothelial receptor, with the exception of CD36.

A functional mutation in codon 125 of the *PECAM-1* gene (leucine → valine replacement in the protein) was analysed in terms of its possible influence on malaria resistance (Casals-Pascual *et al.*, 2001). This study, carried out in Madang (Papua New Guinea) and Kilifi (Kenya) revealed no association between codon 125 polymorphism and disease severity in either of the populations. In another study, carried out in Thailand, a *PECAM-1* haplotype that is significantly associated with CM was reported (Kikuchi *et al.*, 2001).

It seems clear that our knowledge of the role of PECAM-1 in the pathogenicity of malaria is, at present, quite limited. This may be partly explained by the fact that, despite being an apparently high affinity receptor for iRBCs, PECAM-1 has been identified as such more recently than, for example, CD36 and ICAM-1. The use of appropriate animal models and larger-scale assessment of human *PECAM-1* polymorphisms would be a welcome source of potentially valuable information concerning this molecule.

CR1. Complement receptor 1 (CR1; C3b/C4b receptor; CD35), encoded by *CR1*, is a glycoprotein expressed on the surface of human RBCs and leukocytes (see Moulds *et al.*, 1991). CR1 was first implicated in malaria when its involvement in rosette formation, a phenotype associated with disease severity, was shown (Rowe *et al.*, 1997). CR1-dependent rosette formation is common in *P. falciparum* field isolates and the region of CR1 involved in rosetting was mapped (Rowe *et al.*, 2000).

The common blood group antigens, Kn(a)/Kn(b) (Knops); McC(a)/McC(b) (McCoy); Sl(a)/Vil (Swain-Langley/Villien) (now known as Sl:1/Sl:2); and Yk(a) (York) have been shown to be expressed on CR1 and, thus, correspond to alleles that encode polymorphisms in this receptor (Moulds *et al.*, 1991, 2000, 2001). The null phenotypes for the CR1-related blood group (also known as the Knops blood group) antigens are associated with the expression of a low number of CR1 molecules on the erythrocytic surface (Moulds *et al.*, 1991). The Sl(a⁻) and the McC(b⁺) phenotypes had a significantly higher prevalence in a Malian population when compared to other African and European-American populations (Moulds *et al.*, 2000). However, no correlation was found between either the Sl(a⁻) or the McC(b⁺) phenotypes (or, in fact, the *Sl:2/McC(b⁺)* allele) and protection against malaria in The Gambia (Zimmerman *et al.*, 2003).

CR1 is also polymorphic with respect to molecular weight (Wilson and Pearson, 1986; Wong *et al.*, 1989). A *HindIII*-RFLP (restriction fragment length polymorphism), was identified and shown to correlate with differential quantitative expression of CR1 on RBCs in Caucasian populations (Wong *et al.*, 1989). Homozygotes for a 7.4 kilobase (kb) *HindIII* genomic fragment (the H allele) have high RBC CR1 density, whereas homozygotes for a 6.9 kb *HindIII* genomic fragment (the L allele) have low CR1 expression, with HL heterozygotes having intermediate CR1 levels (Wong *et al.*, 1989). Surprisingly, in a Thai population, malaria severity was most prevalent in individuals homozygous for the L allele, compared with heterozygous individuals and individuals homozygous for the H allele (Nagayasu *et al.*, 2001). However, a recent study has found no correlation between the level of erythrocytic CR1 and the H and L alleles in an African population (Rowe *et al.*, 2002), raising doubts about the *HindIII*-RFLP importance in determining CR1 density on RBCs. To add to the confusion, a recent report in Papua New Guinea showed that HH individuals were at most risk of developing SM whereas HL individuals were significantly protected and LL individuals showed statistically insignificant, albeit reduced, odds (Cockburn *et al.*, 2004).

It is clear that further studies are needed in order to fully understand the actual role of CR1-mediated rosetting in determining malaria severity and in establishing a clear correlation between CR1 polymorphisms and protection against severe disease. Studies involving murine models are unlikely to provide any further insights into this issue. In fact, although mouse leukocytes have been shown to contain CR1, it has been shown that mouse RBCs are CR1-negative, in contrast to human RBCs (Rabellino *et al.*, 1978; Kinoshita *et al.*, 1988). Moreover, recent work carried out with the rodent malaria laboratory model *P. chabaudi* showed that rosetting does occur in mice but suggested that the molecules involved may differ from those in human-infecting parasite species (Mackinnon *et al.*, 2002).

The results obtained when studying adhesion molecule polymorphisms in relation to malaria severity were, in all cases described, ambiguous, if not contradictory. This raises intriguing questions regarding *Plasmodium*-driven selection as well as important issues of therapeutic relevance.

3.4 *Host's immune response*

Any infectious disease is characterized, at the host level, by a complex reaction from the host immune system. The counteraction of host invasion by replicating pathogens demands a rapid response, generally provided by components of the innate immune system, which develops promptly and precedes the time-consuming clonal expansion of antigen-specific lymphocytes (Ismail *et al.*, 2002). Recent research has collected evidence of the importance of innate immunity in shaping the subsequent adaptive immune response to malaria blood-stage infection. It is now known that during blood-stage infection there is a 'cross-talk' between the parasite and cells of the innate immune system, such as DC, monocytes/macrophages, natural killer (NK) cells, NKT cells, and γδ T cells. The activity of NK cells was found to be high in uncomplicated cases of malaria while in patients suffering CM there was a profound depression of NK activity (Stach *et al.*, 1986). Murine malaria studies also presented evidence that NK cells play a role in providing protection against the early stages of *P. berghei* or *P. chabaudi* infections (Solomon *et al.*, 1985; Mohan *et al.*, 1997). Recently, using the murine model *P. berghei* ANKA, it was shown that the natural killer complex (NKC) regulates CM, pulmonary edema and severe anaemia, and influences acquired immune responses to infection (Hansen *et al.*, 2005).

Although several recent studies have partially elucidated the role of NK cells during malaria infection (Orago and Facer, 1991; Artavanis-Tsakonas *et al.*, 2003; Artavanis-Tsakonas and Riley, 2002; Baratin *et al.*, 2005), the ligands and receptors responsible for NK-cell activation are still unknown. However, there are some data suggesting an association between NK cell reactivity to *P. falciparum*-infected RBCs and killer Ig-like receptors (KIR) genotype (Artavanis-Tsakonas *et al.*, 2003). This observation raises the fascinating possibility that genetic variation at the *KIR* locus might explain heterogeneity of human NK cell responses to parasitized iRBCs and that the parasite might express ligands to inhibit or activate KIR (Stevenson and Riley, 2004). Nonetheless, these findings highlight the need for large-scale population-based studies in order to address associations between KIR genotype, NK responses and susceptibility to malaria.

The host's immune response to a malaria infection involves not only molecules from the rapid innate immune response but also molecules from a more specific immune response.

IL4. Interleukin-4 (IL4) is a cytokine that regulates the differentiation of precursor T helper cells into the T_H2 subset, which enhance the antigen-presenting capacity of B cells and specific antibody production (Romagnani, 1995). IL4 serves as an important regulator in isotype switching from IgM/IgG to IgE. Several studies have pointed towards IL-4 as an important factor in malaria resistance. A causal relationship between the activation of IL4-producing T-cell subsets and production of the anti-Pf155/RESA-specific antibodies in individuals with immunity induced by a natural malaria infection has been established (Troye-Blomberg *et al.*, 1990). More recently, it was shown that in the Fulani (Burkina Faso), known for having a lower susceptibility to *P. falciparum* infection than their neighbours, the IL4-524T allele (SNP C → T transition at position −524T from the transcription initiation site) was associated with high IgG levels against malaria antigens (Luoni *et al.*, 2001). However, others did not find this association (Verra *et al.*, 2004).

The possible association between three polymorphisms in the IL4 gene with SM in Ghanaian children has been addressed (Gyan *et al.*, 2004). One of these polymorphisms is located in the repeat region (intron3) of the *IL4* gene while the other two are in the promoter region (IL4+33T, SNP at position +33 relative to the transcription initiation site and the previously mentioned IL4-589T). A significantly higher frequency for +33 and −589 loci (IL4+33T/-589T allele) was observed in patients with CM and this was associated with elevated levels of total IgE.

Studies using the murine *P. chabaudi* and *P. berghei* models and IL4-deficient mice failed to show any role for this molecule in parasitaemia control or resistance to infection (von der Weid *et al.*, 1994; Saeftel *et al.*, 2004).

CD40L. CD40 ligand (CD40L) is a glycoprotein expressed in activated T cells. When it binds to CD40 in B cells it regulates their proliferation, antigen-presenting activity and IgG class switching (Durie *et al.*, 1994). CD40L is encoded by the gene *TNFSF5*. The study of CD40L-726 (C/T mutation in the promoter region) and CD40L+220 (C/T mutation in exon 1) polymorphisms association with resistance to SM in *P. falciparum* infections led to the observation that Gambian males with the CD40L-726C allele were protected from CM and severe anaemia (Sabeti *et al.*, 2002). This observation provides evidence to implicate CD40L as a factor in immunity or pathogenesis of malaria infections. The role of CD40L in the course of malaria infections was also explored in *P. berghei* infection, a mouse model of SM (Piguet *et al.*, 2001). CD40L-deficient mice are protected from CM, which seems to occur because the CD40/CD40L system is involved in the breakdown of the blood–brain barrier, macrophage sequestration and platelet consumption.

Fc receptors. The Fc receptors (FcRs) are a family of cell-surface molecules that bind the Fc portion of immunoglobulins. FcRs are widely distributed on cells of the immune system and establish a crucial connection between the humoral and the cellular immune responses (Ravetch and Kinet, 1991). There are three families of FcγR (FcγRI, FcγRII and FcγRIII) that contain multiple distinct genes and alternative splicing forms. Few studies have focused on polymorphisms in different FcγRs. The FcγRIIA presents a polymorphism at position 131 that consists of a single histidine → arginine amino acid substitution (FcγRIIA-131H/R). It was shown that infants with the FcγRIIA-R/R genotype were significantly less likely to be at risk from high-density *P. falciparum* infection, compared with infants with the FcγRIIA-H/R. A later study focused on this same polymorphism and on another receptor polymorphism, FcγRIIIB-NA1/ NA2, known to influence the phagocytic capacity of neutrophils (Omi *et al.*, 2002). This work, performed in northwest Thailand, has shown that malaria severity in this area is not associated with the FcγRIIA-131H/R or the FcγRIIIB-NA1/NA2 poly-morphisms individually. However, the FcγRIIIB-NA2 allele together with the geno-type FcγRIIA-131H/H was shown to be associated with susceptibility to CM. It has been also shown that in West Africa the polymorphism FcγRIIA-131H/R is related to disease severity since the FcγRIIA-131H/H genotype is significantly associated with increased susceptibility to SM (Cooke *et al.*, 2003).

The importance of Fc receptors in a malaria infection was addressed using transgenic mice in two different studies using *P. yoelii yoelii* and *P. berghei* XAT rodent models that showed that host resistance is mediated by antibodies (Rotman *et al.*, 1998; Yoneto *et al.*, 2001). Overall, the different studies indicate that Fc receptors have an important role in malaria infections.

TNF-α. Tumour necrosis factor-α (TNF-α) is a pro-inflammatory cytokine, produced by macrophages, NK cells and T cells, involved in local inflammation and endothelial activation (Janeway, 2001). This cytokine plays a pivotal role in the modulation of immune functions to infection and it was shown to limit the spread of pathogens (reviewed in Beutler and Grau, 1993). In a malaria infection TNF-α production by monocytes/macrophages mainly occurs during the blood stage when RBCs rupture and the schizonts, together with parasite-soluble antigens, are released into the blood-stream (Bate *et al.*, 1988; Kwiatkowski *et al.*, 1989; Taverne *et al.*, 1990a, 1990b; Pichyangkul *et al.*, 1994; Kwiatkowski, 1995).

TNF-α's role in malaria has been extensively studied and both a protective and a pathogenic role for this cytokine have been observed during infection. TNF-α's dual actions are related to its ability to activate endothelial cells, leading to the release of pro-inflammatory cytokines (Pober and Cotran, 1990), to the impairment of the blood–brain barrier (Sharief *et al.*, 1992; Sharief and Thompson, 1992) and to the up-regulation of adhesion molecules (reviewed in Meager, 1999). Of all up-regulated adhesion molecules, TNF-α's interplay with ICAM-1 seems to be extremely impor-tant to malaria pathogenesis (Dietrich, 2002; reviewed in Gimenez *et al.*, 2003). TNF-α also induces nitric oxide (NO) release by endothelial cells (Lamas *et al.*, 1991; Rockett *et al.*, 1992) which seems to play an important role during malaria infections (see NO section below). TNF-α was shown to play a beneficial role by inhibiting *P. falciparum* growth *in vitro* (Haidaris *et al.*, 1983; Rockett *et al.*, 1988). However, this effect was not observed using recombinant TNF (Taverne *et al.*, 1987). TNF was also reported to inhibit infection in murine malaria *in vivo* (Clark *et al.*, 1987; Taverne *et al.*, 1987). These and other later studies suggest that TNF is important in the control of malarial parasites (Ferrante *et al.*, 1990; Kumaratilake *et al.*, 1997). Nevertheless,

TNF-α has also an important role in SM pathogenesis (reviewed in Richards, 1997; Mazier et al., 2000; Odeh, 2001; Gimenez et al., 2003; Clark et al., 2004). High TNF-α levels and malaria severity have been associated in several clinical studies (Grau et al., 1989b; Kwiatkowski et al., 1990; el-Nashar et al., 2002; Esamai et al., 2003) and the ratio of the antagonistic cytokines TNF-α and Interleukin-10 (IL-10) has been shown to be important to the outcome of malaria infection (May et al., 2000) (see IL-10 section). It was shown that the balance between the protective and pathological actions of TNF-α depends on several factors, namely the amount, timing, and duration of TNF-α production, as well as the organ-specific site of synthesis (Beutler and Cerami, 1988; Jacobs et al., 1996b).

TNF-α's unquestionable role in malaria pathogenesis has led to research focused on TNF polymorphisms and their association with malaria severity. Several polymorphisms in the TNF gene, with special emphasis for 2 promoter polymorphism, the TNF-308 and TNF-238, located -308 and -238 nucleotides relative to the gene transcriptional start, have been shown to be important for infection severity (McGuire et al., 1994; Wattavidanage et al., 1999; Aidoo et al., 2001; Wilson et al., 1997; Abraham and Kroeger, 1999; McGuire et al., 1999; Knight et al., 1999; Meyer et al., 2002; Mombo et al., 2003; Flori et al., 2003). While many studies suggest a positive association between the mentioned TNF-α polymorphisms and malaria severity, other studies have failed to find these associations (Ubalee et al., 2001; Stirnadel et al., 1999; Bayley et al., 2004). However, most evidence supports the idea that TNF polymorphisms may be part of the genetic determinants for human malaria resistance/susceptibility. More studies and functional analysis need to be carried out in order to understand the mechanisms involved. Nevertheless the TNF polymorphisms studies performed to date have provided more biological evidence for the role of TNF-α in human malaria.

A different set of studies, following a more therapeutic approach, has assessed the clinical outcome of SM using antibodies to neutralize TNF-α levels (Kwiatkowski et al., 1993; van Hensbroek et al., 1996; Looareesuwan et al., 1999; Jacobs et al., 1996b; Hermsen et al., 1997; Grau et al., 1987) and pentoxifylline to inhibit TNF-α (Sullivan et al., 1988; Strieter et al., 1988; Kremsner et al., 1991; Stoltenburg-Didinger et al., 1993; Di Perri et al., 1995; Hemmer et al., 1997; Looareesuwan et al., 1998; Das et al., 2003) both in human infections and rodent models of infection. A recent report explores a new way to treat CM by inhibiting TNF-α through the use of LMP-420, an anti-inflammatory drug (Wassmer et al., 2005). Data from this report suggest that LMP-420, through the inhibition of endothelial activation, should be considered as a potential way to treat CM. However, the experimental results were based only on in vitro assays and further in vivo experiments are required to assess LMP-420 as a new therapeutic treatment.

While some observations suggest that TNF-α inhibition therapies would be of great value for malaria pathology control, others show that they may not hold any benefit. This lack of consensus may reflect the dual action that TNF-α seems to have during malaria infection.

IFN-γ. Interferon-γ (IFN-γ), mainly produced by NK cells and helper T cells is a pro-inflammatory cytokine involved in macrophage activation, Ig class switching, increased expression of MHC molecules and antigen processing components (Mohan et al., 1997; Janeway et al., 2001; Seixas et al., 2002). IFN-γ plays a central role in the immune response to several infectious diseases (Pfefferkorn and Guyre, 1984; Suzuki et al., 1988; Rossi-Bergmann et al., 1993; van den Broek et al., 1995).

IFN-γ has been shown to be produced during malaria infections (Brake et al., 1988; Meding et al., 1990; Waki et al., 1992; Mohan et al., 1997; Artavanis-Tsakonas et al., 2003; Artavanis-Tsakonas and Riley, 2002; Hailu et al., 2004), and both a protective and a pathological function have been reported. Concerning its protective effect, it was demonstrated that an early IFN-γ-driven Th1-type response is required for an effective control of parasite multiplication in a primary blood-stage infection (Clark, 1987; Shear et al., 1989; Stevenson et al., 1990b; Taylor-Robinson, 1995; Favre et al., 1997; Choudhury et al., 2000; Kobayashi et al., 2000). During infection, IFN-γ activates mono-cytes/macrophages (Bate et al., 1988) and neutrophils (Kumaratilake et al., 1991), which were shown to be involved in the recognition and removal of either merozoites or iRBCs (Khusmith et al., 1982; Kharazmi et al., 1984; Kharazmi and Jepsen, 1984; Ockenhouse et al., 1984; Bouharoun-Tayoun et al., 1995). This protective effect is strongly supported by several studies with rodent models using exogenous IFN-γ (Clark, 1987; Shear et al., 1989) or an anti-IFN-γ mouse antibody (Kobayashi et al., 2000). Furthermore, others have shown a correlation between IFN-γ production and a more positive outcome of malaria infections in both P. chabaudi and P. yoelii (Shear et al., 1989; Stevenson et al., 1990a). Endogenous IFN-γ has also been shown to play a role in the development of pro-tective immunity using IFN-γ and IFN-γ receptor-deficient mice (van der Heyde et al., 1997; Favre et al., 1997). Altogether, these studies with different mouse strains combined with different Plasmodium spp. support a beneficial role for IFN-γ. Several human stud-ies also point toward a protective role for IFN-γ in a malaria infection (Luty et al., 1999; Torre et al., 2001, 2002).

Despite this, several studies have clearly linked IFN-γ to the onset of pathology in mice as well as in humans. The detrimental effects of IFN-γ are thought to be due to its ability to activate macrophages, which, in turn, produce TNF-α, IL-1 and IL-6, leading to activation of an inflammatory cascade (Kern et al., 1989; Day et al., 1999). Grau et al. (1989a) showed that treatment of P. berghei ANKA CBA infected mice with a neutral-izing anti-IFN-γ mouse antibody prevented development of CM and was associated with a significant decrease of TNF-α serum levels. This study supports the idea that there may be a fine balance between the levels of IFN-γ and TNF-α and protective immunity or pathological consequences. Furthermore, a synergy between IFN-γ and TNF-α, particularly with respect to the effects on endothelial cells, has also been observed (Pober et al., 1986). Additional data correlating IFN-γ production with the susceptibility to CM are provided by the observation that CM-susceptible mice exhibit a preferential expan-sion of Th1-like clones characterized by a marked production of IFN-γ (de Kossodo and Grau, 1993). In addition, IFN-γ receptor deficient mice are completely protected from CM (Rudin et al., 1997; Amani et al., 2000). IFN-γ association with malaria pathology in humans is based on the fact that patients with acute P. falciparum malaria present high IFN-γ serum levels (Ringwald et al., 1991; Ho et al., 1995; Wenisch et al., 1995; Nagamine et al., 2003) and individuals at risk for SM produce more IFN-γ in an antigen-specific manner (Chizzolini et al., 1990; Riley et al., 1991).

IFN-γ and its receptor are encoded by the *IFNG* and *IFNGR1* genes, respectively. It was shown that in the Mandika, the predominant Gambian ethnic group, those that are heterozygous for IFNGR1-56 polymorphism (SNP at position -56 (T→C) in the promoter region), presented a two-fold protection against CM and a four-fold protection against death resulting from CM (Koch *et al.*, 2002). Later, it was shown that this polymorphic allele reduces *IFNGR1* gene expression (Juliger *et al.*, 2003).

Only recently has research focused on *IFNG* polymorphisms. SNPs in the region of *IFNG* and the neighbouring *IL22* and *IL26* genes were analysed but no evidence of a strong association between SM and *IFNG* markers was found (Koch *et al.*, 2005). Recently it was shown that in children from Mali the −183T allele is associated with a lower risk of CM (Cabantous *et al.*, 2005). This allele is the polymorphic form of the IFNG-183G/T (G replaced by a T) polymorphism located in the gene promoter region and has been shown to create an AP1-binding site for a nuclear transcription factor, leading to the increase of gene transcription (Bream *et al.*, 2002; Chevillard *et al.*, 2002). IFN-γ clearly plays an important role during malaria infection, which can lead towards protection or pathology. The knowledge of IFN-γ exact involvement in malaria is important for the understanding of infection progression and, consequently, for the assessment of possible therapies against malaria.

NO. Nitric oxide (NO), a labile and highly reactive gas, results from the oxidative deamination of L-arginine to produce L-citruline through a reaction catalysed by the enzyme inducible nitric oxide synthase (iNOS) (Leone *et al.*, 1991). NO has been shown to inhibit the growth and function of diverse infectious agents, such as bacteria, fungi and protozoan parasites, by inactivating some of their critical metabolic pathways (reviewed in James, 1995). Increased NO production has been reported in murine (Taylor-Robinson *et al.*, 1993; Jacobs *et al.*, 1995) and human infections (Kremsner *et al.*, 1996). NO is produced by macrophages (Ahvazi *et al.*, 1995; Tachado *et al.*, 1996) but also by hepatocytes (Mellouk *et al.*, 1991; Nussler *et al.*, 1991) and endothelial cells (Oswald *et al.*, 1994; Tachado *et al.*, 1996). Cytokines such as IFN-γ and TNF-α are critical for the regulation of NO production (Jacobs *et al.*, 1996a). Tachado *et al.* (1996) demonstrated that glycosylphosphatidylinositols (GPI) induce NO release in a time- and dose-dependent manner in macrophages and vascular endothelial cells, and regulate iNOS expression in macrophages. In addition, it was shown that another parasite metabolite, haemozoin, is able to increase IFN-γ-dependent NO production by macrophages (Jaramillo *et al.*, 2003).

Both beneficial and pathological roles have been assigned to NO during malaria infection (Sobolewski *et al.*, 2005a). It is generally accepted that NO kills intraerythrocytic malarial parasites (Brunet, 2001; Clark and Cowden, 2003; Stevenson and Riley, 2004). However, it has been recently proposed that the blood stages are virtually immune to the cytotoxic effects of NO and other reactive oxygen species as a consequence of Hb NO scavenging and reactive oxygen species (ROS) suppression within RBCs (Sobolewski *et al.*, 2005b). Excess of NO can also contribute to malarial immunosuppression (Rockett *et al.*, 1994; Ahvazi *et al.*, 1995) as well as to CM development in *P. falciparum* infections (Clark *et al.*, 1991; Al Yaman *et al.*, 1996). It was proposed that overproduction of NO in the brain might affect local neuronal function by mimicking and exaggerating the physiological effects of endogenous NO (Clark and Cowden, 2003; Clark *et al.*, 1992). However, this hypothesis is not consensual as there is evidence suggesting that NO production may be limited in malaria due to the presence of hypoargininaemia in patients (Lopansri *et al.*, 2003). Furthermore, both plasma and urine NOx levels, as well as the iNOS protein and mRNA levels, correlate inversely with disease severity (Anstey *et al.*, 1996; Boutlis *et al.*, 2003; Perkins *et al.*, 1999; Chiwakata *et al.*, 2000).

A genomic approach, based on the association of *iNOS* gene polymorphisms with malaria infection has been applied to address NO role in malaria. Two SNPs, −954G → C and −1173 C → T, and a pentanucleotide (CCTTT) polymorphism have been studied in terms of malaria clinical outcome. However, reports show controversial data regarding the association of these *iNOS* gene promoter polymorphisms with severe *P. falciparum* malaria. Several reports show that NOS2A-954C allele is associated with protection from SM and resistance to reinfection both in Gabon and Uganda (Kun *et al.*, 1998, 2001; Parikh *et al.*, 2004). However, data from The Gambia, Tanzania and Ghana did not detect this association (Levesque *et al.*, 1999; Burgner *et al.*, 2003; Cramer *et al.*, 2004). A protective role for the *iNOS* -1173 C → T polymorphism has been shown in Tanzania (Hobbs *et al.*, 2002) but not in The Gambia and Ghana (Burgner *et al.*, 2003; Cramer *et al.*, 2004). Both these SNPs were shown to increase NO synthesis (Kun *et al.*, 2001; Hobbs *et al.*, 2002).

A pentanucleotide (CCTTT) microsatellite -2,5 kb of the transcription start site has also been studied. Short microsatellites (<11) have been shown to occur more frequently in fatal CM in The Gambia (Burgner *et al.*, 1998) while longer microsatellite alleles (≥13) were shown to be associated with SM in Thailand and Ghana (Ohashi *et al.*, 2002; Cramer *et al.*, 2004).

These contradictory results may be explained by the fact that the studies reported were carried out in different regions and again by the dual effect that host some molecules seem to have during malaria infection.

IL-10. Interleukin-10 (IL-10) is an anti-inflammatory cytokine produced by T cells and macrophages (Janeway *et al.*, 2001). Expression of IL-10 was shown to increase in response to high TNF-α levels and to down-regulate the production of the latter *in vivo* (Gerard *et al.*, 1993; Howard *et al.*, 1993), possibly representing an attempt of the host to counteract excessive activity of pro-inflammatory cytokines (van der Poll *et al.*, 1994).

Evidence for a beneficial role of IL-10 in malaria stems from an array of different population-based and animal model-based studies. IL-10-deficient mice infected with the rodent model parasite *P. chabaudi chabaudi* have been shown to have higher disease severity and mortality than their wild-type counterparts (Li *et al.*, 1999, 2003; Sanni *et al.*, 2004). The levels of both IL-10 and TNF-α are increased in patients with SM (Day et al., 1999). Several studies have attempted to correlate the IL-10:TNF-α ratio with disease severity (Kurtzhals *et al.*, 1998; Day *et al.*, 1999; Kurtzhals *et al.*, 1999; May *et al.*, 2000; Nussenblatt *et al.*, 2001). The general consensus seems to be that ratios <1 constitute a risk factor for severe anaemia, suggesting that an imbalance between the anti- and the pro-inflammatory responses may constitute a determinant of mortality. Furthermore, a study involving the simian model parasites *P. cynomolgi* and *P. knowlesi*, a role for IL-10 in controlling anaemia during primary infection has been suggested (Praba-Egge *et al.*, 2002).

The available evidence of a protective role for IL-10 against malaria would suggest that strategies to increase the production of this cytokine might hold therapeutic value. However, this may be less simple than it appears. Studies with rodent models have shown that an early pro-inflammatory response may be required to enhance the mechanisms that are essential for elimination of the parasites (Shear *et al.*, 1989; Stevenson *et al.*, 1995; De Souza *et al.*, 1997). In fact, a positive correlation between the IL-10 levels and parasitaemia in *P. falciparum*-infected individuals before the start of anti-malarial treatment has been found (Luty *et al.*, 2000). These results were extended in a recent study that showed a clear association between IL-10 levels and reduced parasite clearance ability in a Tanzanian population undergoing four different treatment regimens with distinct parasite clearance rates (Hugosson *et al.*, 2004). In another study using the rodent parasite model *P. yoelii yoelii*, it was found that mice injected with anti-IL-10 antibody had significantly prolonged survival, suggesting that IL-10 is associated with disease exacerbation rather than protection. Taken together, these observations indicate that enhancing the level of IL-10 might actually be beneficial for the parasite since it could interfere with production of pro-inflammatory cytokines and enable *Plasmodium* to escape effective killing (Hugosson *et al.*, 2004).

Genetic factors substantially influence production of cytokines and may account for as much as 75% of inter-individual differences in IL-10 production (Westendorp *et al.*, 1997). Furthermore, a clear association exists between a variety of IL-10 polymorphisms and susceptibility to several inflammatory diseases, such as rheumatoid arthritis (Eskdale *et al.*, 1998) and systemic lupus erythematosus (Gibson *et al.*, 2001). Thus, the possible correlation between five IL-10 polymorphisms (defining six haplotypes) and malaria severity in a population in The Gambia was investigated (Wilson *et al.*, 2005). One of these haplotypes, termed HAP3, defined by three SNPs at positions +4949, −1117 and −3585, showed a significant association with protection against CM and severe anaemia. However, this case-control association between malaria and HAP3 was not confirmed by a transmission disequilibrium test analysis of the same population. No evidence was found of a correlation between disease severity and any of the individual polymorphisms under study, suggesting that resistance either depends on other SNPs not addressed in this study or on specific combinations of *IL10* polymorphisms (Wilson *et al.*, 2005).

It seems clear that a certain level of IL-10 is required to balance the effect of pro-inflammatory cytokines. Still, an early pro-inflammatory response may be required to enhance the mechanisms that are essential for elimination of the parasites The balance between pro- and anti-inflammatory cytokines during a malaria infection is clearly difficult to ascertain and has important consequences in terms of disease progression. If it is true that fine tuning the inflammatory responses during disease appears as an attractive way to improve its outcome, it is nonetheless obvious that our current knowledge of the complex immunological response that takes place during infection needs to be improved before this can be done in a manner that ensures a promising outcome.

4 Final remarks: exploring the host potential in the 'post-omics era'

The contact between *Plasmodium* and its mammalian host involves a number of interactions that result in a series of different scenarios that range from complete or partial protection to infection and/or disease to SM leading to death. Both sides of this relationship are quite variable and many different possibilities of interaction may occur in the same population. Still, the survival of the *Plasmodium* parasite depends not only on its interaction with host molecules but also on efficiently coping with the host's immune responses. Thus, not only host-driven genetic selection acts on malaria parasite populations, but also *Plasmodium* is likely to exert evolutionary pressure on human gene frequencies. In malaria endemic regions, we might guess that infection contributes positively to the allele frequency of variants associated with protection. The gene conferring the Duffy blood group is one of the most striking examples of this selection pressure; people of this blood type are completely resistant to *P. vivax* blood-stage infection, as they lack a RBC receptor required for *P. vivax* invasion (Hamblin and Di Rienzo, 2000; Chitnis and Miller, 1994). Another striking example is the sickle cell anaemia allele. While severely deleterious in the homozygous state, this allele is associated with malaria protection in the heterozygous state, although the protection mechanisms involved are not fully understood (Aidoo *et al.*, 2002). Importantly, as one might expect, host components involved in the symptomatic phase of infection (blood stage) do not seem to be the only ones implicated in determining the severity of malaria infection. In fact, some alleles on the MHC Class I B53 loci are associated with protective clinical responses in African populations, due to interactions occurring during the silent liver stage of infection (Hill *et al.*, 1991). Presumably this association has contributed to the high frequency of MHC-B53 observed in populations living in areas of high malaria endemicity. Moreover, one should keep in mind that the blood and liver stages of infection can coexist in populations living in endemic areas. Therefore, the final result observed in human populations should not be extrapolated without taking into account possible interactions or common pathways that may occur between the two infection stages.

The sequence of both mice (rodent models of malaria infection) and human genomes can prove to be very useful. Moreover, the use of high-throughput technologies to determine the host molecules altered during the course of an infection, already used for some systems with interesting results (Delahaye *et al.*, 2006), together with systems that allow functional genomics studies, such as RNA interference, will be powerful tools to determine the role of these host molecules during the course of malaria infections.

References

Abraham, L.J. & Kroeger, K.M. (1999) Impact of the -308 TNF promoter polymorphism on the transcriptional regulation of the TNF gene: relevance to disease. *J Leukoc Bio*, **66**: 562–566.

Adams, S., Turner, G.D., Nash, G.B., Micklem, K., Newbold, C.I. & Craig, A.G. (2000) Differential binding of clonal variants of *Plasmodium falciparum* to allelic forms of intracellular adhesion molecule 1 determined by flow adhesion assay. *Infect Immun* **68**: 264–269.

Agarwal, A., Guindo, A., Cissoko, Y., Taylor, J.G., Coulibaly, D., Kone, A., Kayentao, K., Djimde, A., Plowe, C.V., Doumbo, O., Wellems, T.E. & Diallo, D. (2000) Hemoglobin C associated with protection from severe malaria in the Dogon of Mali, a West African population with a low prevalence of hemoglobin S. *Blood* **96**: 2358–2363.

Ahvazi, B.C., Jacobs, P. & Stevenson, M.M. (1995) Role of macrophage-derived nitric oxide in suppression of lymphocyte proliferation during blood-stage malaria. *J Leukoc Biol* **58**: 23–31.

Aidoo, M., McElroy, P.D., Kolczak, M.S., Terlouw, D.J., Ter kuile, F.O., Nahlen, B., Lal, A.A. & Udhayakumar, V. (2001) Tumor necrosis factor-alpha promoter variant 2 (TNF2) is associated with pre-term delivery, infant mortality, and malaria morbidity in western Kenya: Asembo Bay Cohort Project IX. *Genet Epidemiol* **21**: 201–211.

Aidoo, M., Terlouw, D.J., Kolczak, M.S., McElroy, P.D., Ter kuile, F.O., Kariuki, S., Nahlen, B.L., Lal, A.A. & Udhayakumar, V. (2002) Protective effects of the sickle cell gene against malaria morbidity and mortality. *Lancet* **359**: 1311–1312.

Aitman, T.J., Cooper, L.D., Norsworthy, P.J., Wahid, F.N., Gray, J.K., Curtis, B.R., Mckeigue, P.M., Kwiatkowski, D., Greenwood, B.M., Snow, R.W., Hill, A.V. & Scott, J. (2000) Malaria susceptibility and CD36 mutation. *Nature* **405**: 1015–1016.

Al yaman, F.M., Mokela, D., Genton, B., Rockett, K.A., Alpers, M.P. & Clark, I.A. (1996) Association between serum levels of reactive nitrogen intermediates and coma in children with cerebral malaria in Papua New Guinea. *Trans R Soc Trop Med Hyg* **90**: 270–273.

Allen, S.J., O'Donnell, A., Alexander, N.D., Mgone, C.S., Peto, T.E., Clegg, J.B., Alpers, M.P. & Weatherall, D.J. (1999) Prevention of cerebral malaria in children in Papua New Guinea by southeast Asian ovalocytosis band 3. *Am J Trop Med Hyg* **60**: 1056–1060.

Alper, S.L., Darman, R.B., Chernova, M.N. & Dahl, N.K. (2002) The AE gene family of Cl/HCO3- exchangers. *J Nephrol* **15**: suppl 5, s41–s53.

Amani, V., Vigario, A.M., Belnoue, E., Marussig, M., Fonseca, L., Mazier, D. & Renia, L. (2000) Involvement of IFN-gamma receptor-mediated signaling in pathology and anti-malarial immunity induced by *Plasmodium berghei* infection. *Eur J Immunol* **30**: 1646–1655.

Amodu, O.K., Gbadegesin, R.A., Ralph, S.A., Adeyemo, A.A., Brenchley, P.E., Ayoola, O.O., Orimadegun, A.E., Akinsola, A.K., Olumese, P.E. & Omotade, O.O. (2005) *Plasmodium falciparum* malaria in south-west Nigerian children: is the polymorphism of ICAM-1 and E-selectin genes contributing to the clinical severity of malaria? *Acta Trop* **95**: 248–255.

Andrews, K.T. & Lanzer, M. (2002) Maternal malaria: *Plasmodium falciparum* sequestration in the placenta. *Parasitol Res* **88**: 715–723.

Anstey, N.M., Weinberg, J.B., Hassanali, M.Y., Mwaikambo, E.D., Manyenga, D., Misukonis, M.A., Arnelle, D.R., Hollis, D., Mcdonald, M.I. & Granger, D.L. (1996) Nitric oxide in Tanzanian children with malaria: inverse relationship between malaria severity and nitric oxide production/nitric oxide synthase type 2 expression. *J Exp Med* **184**: 557–567.

Arie, T., Fairhurst, R.M., Brittain, N.J., Wellems, T.E. & Dvorak, J.A. (2005) Hemoglobin C modulates the surface topography of *Plasmodium falciparum*-infected erythrocytes. *J Struct Biol* **150**: 163–169.

Artavanis-Tsakonas, K. & Riley, E.M. (2002) Innate immune response to malaria: rapid induction of IFN-gamma from human NK cells by live *Plasmodium falciparum*-infected erythrocytes. *J Immunol* **169**: 2956–2963.

Artavanis-Tsakonas, K., Eleme, K., McQueen, K.L., Cheng, N.W., Parham, P., Davis, D.M. & Riley, E.M. (2003) Activation of a subset of human NK cells upon contact with *Plasmodium falciparum*-infected erythrocytes. *J Immunol* **171**: 5396–5405.

Ayi, K., Turrini, F., Piga, A. & Arese, P. (2004) Enhanced phagocytosis of ring-parasitized mutant erythrocytes: a common mechanism that may explain protection against *falciparum* malaria in sickle trait and beta-thalassemia trait. *Blood* 104: 3364–3371.

Baratin, M., Roetynck, S., Lepolard, C., Falk, C., Sawadogo, S., Uematsu, S., Akira, S., Ryffel, B., Tiraby, J.G., Alexopoulou, L., Kirschning, C.J., Gysin, J., Vivier, E. & Ugolini, S. (2005) Natural killer cell and macrophage cooperation in MyD88-dependent innate responses to *Plasmodium falciparum*. *Proc Natl Acad Sci USA* 102: 14747–14752.

Barnwell, J.W., Ockenhouse, C.F. & Knowles, D.M., 2nd (1985) Monoclonal antibody OKM5 inhibits the *in vitro* binding of *Plasmodium falciparum*-infected erythrocytes to monocytes, endothelial, and C32 melanoma cells. *J Immunol* 135: 3494–3497.

Barnwell, J.W., Asch, A.S., Nachman, R.L., Yamaya, M., Aikawa, M. & Ingravallo, P. (1989) A human 88-kD membrane glycoprotein (CD36) functions *in vitro* as a receptor for a cytoadherence ligand on *Plasmodium falciparum*-infected erythrocytes. *J Clin Invest* 84: 765–772.

Baruch, D.I., Ma, X.C., Singh, H.B., Bi, X, Pasloske, B.L. & Howard, R.J. (1997) Identification of a region of PfEMP1 that mediates adherence of *Plasmodium falciparum* infected erythrocytes to CD36: conserved function with variant sequence. *Blood* 90: 3766–3775.

Bate, C.A., Taverne, J. & Playfair, J.H. (1988) Malarial parasites induce TNF production by macrophages. *Immunology* 64: 227–231.

Baum, J., Pinder, M. & Conway, D.J. (2003) Erythrocyte invasion phenotypes of *Plasmodium falciparum* in The Gambia. *Infect Immun* 71: 1856–1863.

Bayley, J.P., Ottenhoff, T.H. & Verweij, C.L. (2004) Is there a future for TNF promoter polymorphisms? *Genes Immun* 5: 315–329.

Beeson, J.G. & Brown, G.V. (2002) Pathogenesis of *Plasmodium falciparum* malaria: the roles of parasite adhesion and antigenic variation. *Cell Mol Life Sci* 59: 258–271.

Beeson, J.G., Reeder, J.C., Rogerson, S.J. & Brown, G.V. (2001) Parasite adhesion and immune evasion in placental malaria. *Trends Parasitol* 17: 331–337.

Beeson, J.G., Rogerson, S.J. & Brown, G.V. (2002) Evaluating specific adhesion of *Plasmodium falciparum*-infected erythrocytes to immobilized hyaluronic acid with comparison to binding of mammalian cells. *Int J Parasitol* 32: 1245–1252.

Bellamy, R., Kwiatkowski, D. & Hill, A.V. (1998) Absence of an association between intercellular adhesion molecule 1, complement receptor 1 and interleukin 1 receptor antagonist gene polymorphisms and severe malaria in a West African population. *Trans R Soc Trop Med Hyg* 92: 312–316.

Berendt, A.R., Simmons, D.L., Tansey, J., Newbold, C.I. & Marsh, K. (1989) Intercellular adhesion molecule-1 is an endothelial cell adhesion receptor for *Plasmodium falciparum*. *Nature* 341: 57–59.

Beutler, B. & Cerami, A. (1988) Tumor necrosis, cachexia, shock, and inflammation: a common mediator. *Annu Rev Biochem* 57: 505–518.

Beutler, B. & Grau, G.E. (1993) Tumor necrosis factor in the pathogenesis of infectious diseases. *Crit Care Med* 21: S423–S435.

Beutler, E. (1990) The genetics of glucose-6-phosphate dehydrogenase deficiency. *Semin Hematol* 27: 137–164.

Beutler, E. (1996) G6PD: population genetics and clinical manifestations. *Blood Rev* 10: 45–52.

Bodescot, M., Silvie, O., Siau, A., Refour, P., Pino, P., Franetich, J.F., Hannoun, L., Sauerwein, R. & Mazier, D. (2004) Transcription status of vaccine candidate genes of *Plasmodium falciparum* during the hepatic phase of its life cycle. *Parasitol Res* 92: 449–452.

Bouharoun-Tayoun, H., Oeuvray, C., Lunel, F. & Druilhe, P. (1995) Mechanisms underlying the monocyte-mediated antibody-dependent killing of *Plasmodium falciparum* asexual blood stages. *J Exp Med* 182: 409–418.

Boutlis, C.S., Tjitra, E., Maniboey, H., Misukonis, M.A., Saunders, J.R., Suprianto, S., Weinberg, J.B. & Anstey, N.M. (2003) Nitric oxide production and mononuclear cell nitric oxide synthase activity in malaria-tolerant Papuan adults. *Infect Immun* 71: 3682–3689.

Brake, D.A., Long, C.A. & Weidanz, W.P. (1988) Adoptive protection against *Plasmodium chabaudi adami* malaria in athymic nude mice by a cloned T cell line. *J Immunol* 140: 1989–1993.

Bream, J.H., Ping, A., Zhang, X., Winkler, C. & Young, H.A. (2002) A single nucleotide polymorphism in the proximal IFN-gamma promoter alters control of gene transcription. *Genes Immun* 3: 165–169.

Brittenham, G.M., Schechter, A.N. & Noguchi, C.T. (1985) Hemoglobin S polymerization: primary determinant of the hemolytic and clinical severity of the sickling syndromes. *Blood* 65: 183–189.

Bruna-Romero, O. & Rodriguez, A. (2001) Dendritic cells can initiate protective immune responses against malaria. *Infect Immun* 69: 5173–5176.

Brunet, L.R. (2001) Nitric oxide in parasitic infections. *Int Immunopharmacol* 1: 1457–1467.

Burgner, D., Xu, W., Rockett, K., Gravenor, M., Charles, I.G., Hill, A.V. & Kwiatkowski, D. (1998) Inducible nitric oxide synthase polymorphism and fatal cerebral malaria. *Lancet* 352: 1193–1194.

Burgner, D., Usen, S., Rockett, K., Jallow, M., Ackerman, H., Cervino, A., Pinder, M. & Kwiatkowski, D.P. (2003) Nucleotide and haplotypic diversity of the NOS2A promoter region and its relationship to cerebral malaria. *Hum Genet* 112: 379–386.

Cabantous, S., Poudiougou, B., Traore, A., Keita, M., Cisse, M.B., Doumbo, O., Dessein, A.J. & Marquet, S. (2005) Evidence that interferon-gamma plays a protective role during cerebral malaria. *J Infect Dis* 192: 854–860.

Cappadoro, M., Giribaldi, G., O'Brien, E., Turrini, F., Mannu, F., Ulliers, D., Simula, G., Luzzatto, L. & Arese, P. (1998) Early phagocytosis of glucose-6-phosphate dehydrogenase (G6PD)-deficient erythrocytes parasitized by *Plasmodium falciparum* may explain malaria protection in G6PD deficiency. *Blood* 92: 2527–2534.

Carrolo, M., Giordano, S., Cabrita-Santos, L., Corso, S., Vigario, A.M., Silva, S., Leiriao, P., Carapau, D., Armas-Portela, R., Comoglio, P.M., Rodriguez, A. & Mota, M.M. (2003) Hepatocyte growth factor and its receptor are required for malaria infection. *Nat Med* 9: 1363–1369.

Carvalho, L.J., Daniel-Ribeiro, C.T. & Goto, H. (2002) Malaria vaccine: candidate antigens, mechanisms, constraints and prospects. *Scand J Immunol* 56: 327–343.

Casals-Pascual, C., Allen, S., Allen, A., Kai, O., Lowe, B., Pain, A. & Roberts, D.J. (2001) Short report: codon 125 polymorphism of CD31 and susceptibility to malaria. *Am J Trop Med Hyg* 65: 736–737.

Chevillard, C., Henri, S., Stefani, F., Parzy, D. & Dessein, A. (2002) Two new polymorphisms in the human interferon gamma (IFN-gamma) promoter. *Eur J Immunogenet* **29**: 53–56.

Chitnis, C.E. & Miller, L.H. (1994) Identification of the erythrocyte binding domains of *Plasmodium vivax* and *Plasmodium knowlesi* proteins involved in erythrocyte invasion. *J Exp Med* **180**: 497–506.

Chiwakata, C.B., Hemmer, C.J. & Dietrich, M. (2000) High levels of inducible nitric oxide synthase mRNA are associated with increased monocyte counts in blood and have a beneficial role in *Plasmodium falciparum* malaria. *Infect Immun* **68**: 394–399.

Chizzolini, C., Grau, G.E., Geinoz, A. & Schrijvers, D. (1990) T lymphocyte interferon-gamma production induced by *Plasmodium falciparum* antigen is high in recently infected non-immune and low in immune subjects. *Clin Exp Immunol* **79**: 95–99.

Chotivanich, K., Udomsangpetch, R., Pattanapanyasat, K., Chierakul, W., Simpson, J., Looareesuwan, S. & White, N. (2002) Hemoglobin E: a balanced polymorphism protective against high parasitemias and thus severe *P. falciparum* malaria. *Blood* **100**: 1172–1176.

Choudhury, H.R., Sheikh, N.A., Bancroft, G.J., Katz, D.R. & de Souza, J.B. (2000) Early nonspecific immune responses and immunity to blood-stage nonlethal *Plasmodium yoelii* malaria. *Infect Immun* **68**: 6127–6132.

Clark, I.A. (1987) Cell-mediated immunity in protection and pathology of malaria. *Parasitol Today* **3**: 300–305.

Clark, I.A. & Cowden, W.B. (2003) The pathophysiology of *falciparum* malaria. *Pharmacol Ther* **99**: 221–260.

Clark, I.A., Hunt, N.H., Butcher, G.A. & Cowden, W.B. (1987) Inhibition of murine malaria (*Plasmodium chabaudi*) *in vivo* by recombinant interferon-gamma or tumor necrosis factor, and its enhancement by butylated hydroxyanisole. *J Immunol* **139**: 3493–3496.

Clark, I.A., Rockett, K.A. & Cowden, W.B. (1991) Proposed link between cytokines, nitric oxide and human cerebral malaria. *Parasitol Today* **7**: 205–207.

Clark, I.A., Rockett, K.A. & Cowden, W.B. (1992) Possible central role of nitric oxide in conditions clinically similar to cerebral malaria. *Lancet* **340**: 894–896.

Clark, I.A., Alleva, L.M., Mills, A.C. & Cowden, W.B. (2004) Pathogenesis of malaria and clinically similar conditions. *Clin Microbiol Rev* **17**: 509–539, table of contents.

Clyde, D.F., Most, H., McCarthy, V.C. & Vanderberg, J.P. (1973) Immunization of man against sporozite-induced *falciparum* malaria. *Am J Med Sci* **266**: 169–177.

Cockburn, I.A., MacKinnon, M.J., O'Donnell, A., Allen, S.J., Moulds, J.M., Baisor, M., Bockarie, M., Reeder, J.C. & Rowe, J.A. (2004) A human complement receptor 1 polymorphism that reduces *Plasmodium falciparum* rosetting confers protection against severe malaria. *Proc Natl Acad Sci USA* **101**: 272–277.

Cooke, B.M., Nicoll, C.L., Baruch, D.I. & Coppel, R.L. (1998) A recombinant peptide based on PfEMP-1 blocks and reverses adhesion of malaria-infected red blood cells to CD36 under flow. *Mol Microbiol* **30**: 83–90.

Cooke, G.S., Aucan, C., Walley, A.J., Segal, S., Greenwood, B.M., Kwiatkowski, D.P. & Hill, W.V. (2003) Association of Fcgamma receptor IIa (CD32) polymorphism with severe malaria in West Africa. *Am J Trop Med Hyg* **69**: 565–568.

Craig, A., Fernandez-Reyes, D., Mesri, M., McDowall, A., Altieri, D.C., Hogg, N. & Newbold, C. (2000) A functional analysis of a natural variant of intercellular adhesion molecule-1 (ICAM-1Kilifi). *Hum Mol Genet* **9**: 525–530.

Cramer, J.P., Mockenhaupt, F.P., Ehrhardt, S., Burkhardt, J., Otchwemah, R.N., Dietz, E., Gellert, S. & Bienzle, U. (2004) iNOS promoter variants and severe malaria in Ghanaian children. *Trop Med Int Health* **9**: 1074–1080.

Cranston, H.A., Boylan, C.W., Carroll, G.L., Sutera, S.P., Williamson, J.R., Gluzman, I.Y. & Krogstad, D.J. (1984) *Plasmodium falciparum* maturation abolishes physiologic red cell deformability. *Science* **223**: 400–403.

Das, B.K., Mishra, S., Padhi, P.K., Manish, R., Tripathy, R., Sahoo, P.K. & Ravindran, B. (2003) Pentoxifylline adjunct improves prognosis of human cerebral malaria in adults. *Trop Med Int Health* **8**: 680–684.

Day, N.P., Hien, T.T., Schollaardt, T., Loc, P.P., Chuong, L.V., Chau, T.T., Mai, N.T., Phu, N.H., Sinh, D.X., White, N.J. & Ho, M. (1999) The prognostic and pathophysiologic role of pro- and antiinflammatory cytokines in severe malaria. *J Infect Dis* **180**: 1288–1297.

De Kossodo, S. & Grau, G.E. (1993) Profiles of cytokine production in relation with susceptibility to cerebral malaria. *J Immunol* **151**: 4811–4820.

De Souza, J.B., Williamson, K.H., Otani, T. & Playfair, J.H. (1997) Early gamma interferon responses in lethal and nonlethal murine blood-stage malaria. *Infect Immun* **65**: 1593–1598.

Delahaye, N.F., Coltel, N., Puthier, D., Flori, L., Houlgatte, R., Iraqi, F.A., Nguyen, C., Grau, G.E. & Rihet, P. (2006) Gene-expression profiling discriminates between cerebral malaria (CM)-susceptible mice and CM-resistant mice. *J Infect Dis* **193**: 312–321.

Di perri, G., Di perri, I.G., Monteiro, G.B., Bonora, S., Hennig, C., Cassatella, M., Micciolo, R., Vento, S., Dusi, S., Bassetti, D. *et al.* (1995) Pentoxifylline as a supportive agent in the treatment of cerebral malaria in children. *J Infect Dis* **171**: 1317–1322.

Dietrich, I.B. (2002) The adhesion molecule ICAM-1 and its regulation in relation with the blood–brain barrier. *J Neuroimmunol* **128**: 58–68.

Dolan, S.A., Miller, L.H. & Wellems, T.E. (1990) Evidence for a switching mechanism in the invasion of erythrocytes by *Plasmodium falciparum*. *J Clin Invest* **86**: 618–624.

Dolan, S.A., Proctor, J.L., Alling, D.W., Okubo, Y., Wellems, T.E. & Miller, L.H. (1994) Glycophorin B as an EBA-175 independent *Plasmodium falciparum* receptor of human erythrocytes. *Mol Biochem Parasitol* **64**: 55–63.

Duffy, P.E. & Fried, M. (2003) *Plasmodium falciparum* adhesion in the placenta. *Curr Opin Microbiol* **6**: 371–376.

Duraisingh, M.T., Maier, A.G., Triglia, T. & Cowman, A.F. (2003) Erythrocyte-binding antigen 175 mediates invasion in *Plasmodium falciparum* utilizing sialic acid-dependent and -independent pathways. *Proc Natl Acad Sci USA* **100**: 4796–4801.

Durie, F.H., Foy, T.M., Masters, S.R., Laman, J.D. & Noelle, R.J. (1994) The role of CD40 in the regulation of humoral and cell-mediated immunity. *Immunol Today* **15**: 406–411.

El-nashar, T.M., El-kholy, H.M., El-shiety, A.G. & Al-zahaby, A.A. (2002) Correlation of plasma levels of tumor necrosis factor, interleukin-6 and nitric oxide with the severity of human malaria. *J Egypt Soc Parasitol* **32**: 525–535.

Esamai, F., Ernerudh, J., Janols, H., Welin, S., Ekerfelt, C., Mining, S. & Forsberg, P. (2003) Cerebral malaria in children: serum and cerebrospinal fluid TNF-alpha and TGF-beta levels and their relationship to clinical outcome. *J Trop Pediatr* 49: 216–223.

Eskdale, J., McNicholl, J., Wordsworth, P., Jonas, B., Huizinga, T., Field, M. & Gallagher, G. (1998) Interleukin-10 microsatellite polymorphisms and IL-10 locus alleles in rheumatoid arthritis susceptibility. *Lancet* 352: 1282–1283.

Esser, M.T., Marchese, R.D., Kierstead, L.S., Tussey, L.G., Wang, F., Chirmule, N. & Washabaugh, M.W. (2003) Memory T cells and vaccines. *Vaccine* 21: 419–430.

Fairhurst, R.M., Baruch, D.I., Brittain, N.J., Ostera, G.R., Wallach, J.S., Hoang, H.L., Hayton, K., Guindo, A., Makobongo, M.O., Schwartz, O.M., Tounkara, A., Doumbo, O.K., Diallo, D.A., Fujioka, H., Ho, M. & Wellems, T.E. (2005) Abnormal display of PfEMP-1 on erythrocytes carrying haemoglobin C may protect against malaria. *Nature* 435: 1117–1121.

Favre, N., Ryffel, B., Bordmann, G. & Rudin, W. (1997) The course of *Plasmodium chabaudi chabaudi* infections in interferon-gamma receptor deficient mice. *Parasite Immunol* 19: 375–383.

Fernandez-Reyes, D., Craig, A.G., Kyes, S.A., Peshu, N., Snow, R.W., Berendt, A.R., Marsh, K. & Newbold, C.I. (1997) A high frequency African coding polymorphism in the N-terminal domain of ICAM-1 predisposing to cerebral malaria in Kenya. *Hum Mol Genet* 6: 1357–1360.

Ferrante, A., Kumaratilake, L., Rzepczyk, C.M. & Dayer, J.M. (1990) Killing of *Plasmodium falciparum* by cytokine activated effector cells (neutrophils and macrophages). *Immunol Lett* 25: 179–187.

Flori, L., Sawadogo, S., Esnault, C., Delahaye, N.F., Fumoux, F. & Rihet, P. (2003) Linkage of mild malaria to the major histocompatibility complex in families living in Burkina Faso. *Hum Mol Genet* 12: 375–378.

Franke-Fayard, B., Janse, C.J., Cunha-Rodrigues, M., Ramesar, J., Buscher, P., Que, I., Lowik, C., Voshol, P.J., den Boer, M.A., van Duinen, S.G., Febbraio, M., Mota, M.M. & Waters, A.P. (2005) Murine malaria parasite sequestration: CD36 is the major receptor, but cerebral pathology is unlinked to sequestration. *Proc Natl Acad Sci USA* 102: 11468–11473.

Frevert, U., Engelmann, S., Zougbede, S., Stange, J., Ng, B., Matuschewski, K., Liebes, L. & Yee, H. (2005) Intravital observation of *Plasmodium berghei* sporozoite infection of the liver. *Plos Biol* 3: e192.

Genton, B., Al-yaman, F., Mgone, C.S., Alexander, N., Paniu, M.M., Alpers, M.P. & Mokela, D. (1995) Ovalocytosis and cerebral malaria. *Nature* 378: 564–565.

Gerard, C., Bruyns, C., Marchant, A., Abramowicz, D., Vandenabeele, P., Delvaux, A., Fiers, W., Goldman, M. & Velu, T. (1993) Interleukin 10 reduces the release of tumor necrosis factor and prevents lethality in experimental endotoxemia. *J Exp Med* 177: 547–550.

Gibson, A.W., Edberg, J.C., Wu, J., Westendorp, R.G., Huizinga, T.W. & Kimberly, R.P. (2001) Novel single nucleotide polymorphisms in the distal IL-10 promoter affect IL-10 production and enhance the risk of systemic lupus erythematosus. *J Immunol* 166: 3915–3922.

Gilbert, S.C., Plebanski, M., Gupta, S., Morris, J., Cox, M., Aidoo, M., Kwiatkowski, D., Greenwood, B.M., Whittle, H.C. & Hill, A.V. (1998)

Association of malaria parasite population structure, HLA, and immunological antagonism. *Science* **279**: 1173–1177.

Gimenez, F., Barraud de Lagerie, S., Fernandez, C., Pino, P. & Mazier, D. (2003) Tumor necrosis factor alpha in the pathogenesis of cerebral malaria. *Cell Mol Life Sci* **60**: 1623–1635.

Grau, G.E., Fajardo, L.F., Piguet, P.F., Allet, B., Lambert, P.H. & Vassalli, P. (1987) Tumor necrosis factor (cachectin) as an essential mediator in murine cerebral malaria. *Science* **237**: 1210–1212.

Grau, G.E., Heremans, H., Piguet, P.F., Pointaire, P., Lambert, P.H., Billiau, A. & Vassalli, P. (1989a) Monoclonal antibody against interferon gamma can prevent experimental cerebral malaria and its associated overproduction of tumor necrosis factor. *Proc Natl Acad Sci USA* **86**: 5572–5574.

Grau, G.E., Taylor, T.E., Molyneux, M.E., Wirima, J.J., Vassalli, P., Hommel, M. & Lambert, P.H. (1989b) Tumor necrosis factor and disease severity in children with *falciparum* malaria. *N Engl J Med* **320**: 1586–1591.

Gruner, A.C., Snounou, G., Brahimi, K., Letourneur, F., Renia, L. & Druilhe, P. (2003) Pre-erythrocytic antigens of *Plasmodium falciparum*: from rags to riches? *Trends Parasitol* **19**: 74–78.

Gyan, B.A., Goka, B., Cvetkovic, J.T., Kurtzhals, J.L., Adabayeri, V., Perlmann, H., Lefvert, A.K., Akanmori, B.D. & Troye-Blomberg, M. (2004) Allelic polymorphisms in the repeat and promoter regions of the interleukin-4 gene and malaria severity in Ghanaian children. *Clin Exp Immunol* **138**: 145–150.

Hadley, T.J., Klotz, F.W., Pasvol, G., Haynes, J.D., McGinniss, M.H., Okubo, Y. & Miller, L.H. (1987) *Falciparum* malaria parasites invade erythrocytes that lack glycophorin A and B (MkMk). Strain differences indicate receptor heterogeneity and two pathways for invasion. *J Clin Invest* **80**: 1190–1193.

Haidaris, C.G., Haynes, J.D., Meltzer, M.S. & Allison, A.C. (1983) Serum containing tumor necrosis factor is cytotoxic for the human malaria parasite *Plasmodium falciparum*. *Infect Immun* **42**: 385–393.

Hailu, A., van der Poll, T., Berhe, N. & Kager, P.A. (2004) Elevated plasma levels of interferon (IFN)-gamma, IFN-gamma inducing cytokines, and IFN-gamma inducible CXC chemokines in visceral leishmaniasis. *Am J Trop Med Hyg* **71**: 561–567.

Hamblin, M.T. & Di Rienzo, A. (2000) Detection of the signature of natural selection in humans: evidence from the Duffy blood group locus. *Am J Hum Genet* **66**: 1669–1679.

Hamblin, M.T., Thompson, E.E. & Di Rienzo, A. (2002) Complex signatures of natural selection at the Duffy blood group locus. *Am J Hum Genet* **70**: 369–383.

Hansen, D.S., Evans, K.J., D'Ombrain, M.C., Bernard, N.J., Sexton, A.C., Buckingham, L., Scalzo, A.A. & Schofield, L. (2005) The natural killer complex regulates severe malarial pathogenesis and influences acquired immune responses to *Plasmodium berghei* ANKA. *Infect Immun* **73**:2288–2297.

Heddini, A., Chen, Q., Obiero, J., Kai, O., Fernandez, V., Marsh, K., Muller, W.A. & Wahlgren, M. (2001a) Binding of *Plasmodium falciparum*-infected erythrocytes to soluble platelet endothelial cell adhesion molecule-1 (PECAM-1/CD31): frequent recognition by clinical isolates. *Am J Trop Med Hyg* **65**: 47–51.

Heddini, A., Pettersson, F., Kai, O., Shafi, J., Obiero, J., Chen, Q., Barragan, A., Wahlgren, M. & Marsh, K. (2001b) Fresh isolates from children with severe *Plasmodium falciparum* malaria bind to multiple receptors. *Infect Immun* **69**: 5849–5856.

Hemmer, C.J., Hort, G., Chiwakata, C.B., Seitz, R., Egbring, R., Gaus, W., Hogel, J., Hassemer, M., Nawroth, P.P., Kern, P. & Dietrich, M. (1997) Supportive pentoxifylline in *falciparum* malaria: no effect on tumor necrosis factor alpha levels or clinical outcome: a prospective, randomized, placebo-controlled study. *Am J Trop Med Hyg* **56**: 397–403.

Hermsen, C.C., Crommert, J.V., Fredix, H., Sauerwein, R.W. & Eling, W.M. (1997) Circulating tumour necrosis factor alpha is not involved in the development of cerebral malaria in *Plasmodium berghei*-infected c57bl mice. *Parasite Immunol* **19**: 571–577.

Herrington, D., Davis, J., Nardin, E., Beier, M., Cortese, J., Eddy, H., Losonsky, G., Hollingdale, M., Sztein, M., Levine, M., *et al.* (1991) Successful immunization of humans with irradiated malaria sporozoites: humoral and cellular responses of the protected individuals. *Am J Trop Med Hyg* **45**: 539–547.

Hill, A.V. (2001) The genomics and genetics of human infectious disease susceptibility. *Annu Rev Genomics Hum Genet* **2**: 373–400.

Hill, A.V., Allsopp, C.E., Kwiatkowski, D., Anstey, N.M., Twumasi, P., Rowe, P.A., Bennett, S., Brewster, D., McMichael, A.J. & Greenwood, B.M. (1991) Common West African HLA antigens are associated with protection from severe malaria. *Nature* **352**: 595–600.

Hill, A.V., Elvin, J., Willis, A.C., Aidoo, M., Allsopp, C.E., Gotch, F.M., Gao, X.M., Takiguchi, M., Greenwood, B.M., Townsend, A.R. *et al.* (1992) Molecular analysis of the association of HLA-B53 and resistance to severe malaria. *Nature* **360**: 434–439.

Ho, M. & White, N.J. (1999) Molecular mechanisms of cytoadherence in malaria. *Am J Physiol* **276**: c1231–c1242.

Ho, M., Sexton, M.M., Tongtawe, P., Looareesuwan, S., Suntharasamai, P. & Webster, H.K. (1995) Interleukin-10 inhibits tumor necrosis factor production but not antigen-specific lymphoproliferation in acute *Plasmodium falciparum* malaria. *J Infect Dis* **172**: 838–844.

Ho, M., Hickey, M.J., Murray, A.G., Andonegui, G. & Kubes, P. (2000) Visualization of *Plasmodium falciparum*-endothelium interactions in human microvasculature: mimicry of leukocyte recruitment. *J Exp Med* **192**: 1205–1211.

Hobbs, M.R., Udhayakumar, V., Levesque, M.C., Booth, J., Roberts, J.M., Tkachuk, A.N., Pole, A., Coon, H., Kariuki, S., Nahlen, B.L., Mwaikambo, E.D., Lal, A.L., Granger, D.L., Anstey, N.M. & Weinberg, J.B. (2002) A new NOS2 promoter polymorphism associated with increased nitric oxide production and protection from severe malaria in Tanzanian and Kenyan children. *Lancet* **360**: 1468–1475.

Howard, M., Muchamuel, T., Andrade, S. & Menon, S. (1993) Interleukin 10 protects mice from lethal endotoxemia. *J Exp Med* **177**: 1205–1208.

Hugosson, E., Montgomery, S.M., Premji, Z., Troye-Blomberg, M. & Bjorkman, A. (2004) Higher IL-10 levels are associated with less effective clearance of *Plasmodium falciparum* parasites. *Parasite Immunol* **26**: 111–117.

Ishino, T., Yano, K., Chinzei, Y. & Yuda, M. (2004) Cell-passage activity is required for the malarial parasite to cross the liver sinusoidal cell layer. *Plos Biol* 2: e4.

Ismail, N., Olano, J.P., Feng, H.M. & Walker, D.H. (2002) Current status of immune mechanisms of killing of intracellular microorganisms. *Fems Microbiol Lett* 207: 111–120.

Jacobs, P., Radzioch, D. & Stevenson, M.M. (1995) Nitric oxide expression in the spleen, but not in the liver, correlates with resistance to blood-stage malaria in mice. *J Immunol* 155: 5306–5313.

Jacobs, P., Radzioch, D. & Stevenson, M.M. (1996a) *In vivo* regulation of nitric oxide production by tumor necrosis factor alpha and gamma interferon, but not by interleukin-4, during blood stage malaria in mice. *Infect Immun* 64: 44–49.

Jacobs, P., Radzioch, D. & Stevenson, M.M. (1996b) A Th1-associated increase in tumor necrosis factor alpha expression in the spleen correlates with resistance to blood-stage malaria in mice. *Infect Immun* 64: 535–541.

James, S.L. (1995) Role of nitric oxide in parasitic infections. *Microbiol Rev* 59: 533–547.

Janeway C.A, T.P., Walport M, Shlomchik M (2001) Cytokines and their receptors. *Immunobiology: The Immune System in Health and Disease*, 5 edn. Garland Science Textbooks, Taylor and Francis, New York and London.

Jaramillo, M., Gowda, D.C., Radzioch, D. & Olivier, M. (2003) Hemozoin increases IFN-gamma-inducible macrophage nitric oxide generation through extracellular signal-regulated kinase- and NF-kappa B-dependent pathways. *J Immunol* 171: 4243–4253.

Juliger, S., Bongartz, M., Luty, A.J., Kremsner, P.G. & Kun, J.F. (2003) Functional analysis of a promoter variant of the gene encoding the interferon-gamma receptor chain I. *Immunogenetics* 54: 675–680.

Jung, S., Unutmaz, D., Wong, P., Sano, G., De Los Santos, K., Sparwasser, T., Wu, S., Vuthoori, S., Ko, K., Zavala, F., Pamer, E.G., Littman, D.R. & Lang, R.A. (2002) *In vivo* depletion of CD11c(+) dendritic cells abrogates priming of CD8(+) t cells by exogenous cell-associated antigens. *Immunity* 17: 211–220.

Kaul, D.K., Liu, X.D., Nagel, R.L. & Shear, H.L. (1998) Microvascular hemodynamics and *in vivo* evidence for the role of intercellular adhesion molecule-1 in the sequestration of infected red blood cells in a mouse model of lethal malaria. *Am J Trop Med Hyg* 58: 240–247.

Kern, P., Hemmer, C.J., van Damme, J., Gruss, H.J. & Dietrich, M. (1989) Elevated tumor necrosis factor alpha and interleukin-6 serum levels as markers for complicated *Plasmodium falciparum* malaria. *Am J Med* 87: 139–143.

Kharazmi, A. & Jepsen, S. (1984) Enhanced inhibition of *in vitro* multiplication of *Plasmodium falciparum* by stimulated human polymorphonuclear leucocytes. *Clin Exp Immunol* 57: 287–292.

Kharazmi, A., Jepsen, S. & Valerius, N.H. (1984) Polymorphonuclear leucocytes defective in oxidative metabolism inhibit *in vitro* growth of *Plasmodium falciparum*. Evidence against an oxygen-dependent mechanism. *Scand J Immunol* 20: 93–96.

Khusmith, S., Druilhe, P. & Gentilini, M. (1982) Enhanced *Plasmodium falciparum* merozoite phagocytosis by monocytes from immune individuals. *Infect Immun* 35: 874–879.

Kidson, C., Lamont, G., Saul, A. & Nurse, G.T. (1981) Ovalocytic erythrocytes from Melanesians are resistant to invasion by malaria parasites in culture. *Proc Natl Acad Sci USA* **78**: 5829–5832.

Kikuchi, M., Looareesuwan, S., Ubalee, R., Tasanor, O., Suzuki, F., Wattanagoon, Y., Na-Bangchang, K., Kimura, A., Aikawa, M. & Hirayama, K. (2001) Association of adhesion molecule PECAM-1/CD31 polymorphism with susceptibility to cerebral malaria in Thais. *Parasitol Int* **50**: 235–239.

Kinoshita, T., Takeda, J., Hong, K., Kozono, H., Sakai, H. & Inoue, K. (1988) Monoclonal antibodies to mouse complement receptor type 1 (CR1). Their use in a distribution study showing that mouse erythrocytes and platelets are CR1-negative. *J Immunol* **140**: 3066–3072.

Knight, J.C., Udalova, I., Hill, A.V., Greenwood, B.M., Peshu, N., Marsh, K. & Kwiatkowski, D. (1999) A polymorphism that affects OCT-1 binding to the TNF promoter region is associated with severe malaria. *Nat Genet* **22**: 145–150.

Kobayashi, F., Ishida, H., Matsui, T. & Tsuji, M. (2000) Effects of *in vivo* administration of anti-IL:-10 or anti-IFN-gamma monoclonal antibody on the host defense mechanism against *Plasmodium yoelii yoelii* infection. *J Vet Med Sci* **62**: 583–587.

Koch, O., Awomoyi, A., Usen, S., Jallow, M., Richardson, A., Hull, J., Pinder, M., Newport, M. & Kwiatkowski, D. (2002) IFNGR1 gene promoter polymorphisms and susceptibility to cerebral malaria. *J Infect Dis* **185**: 1684–1687.

Koch, O., Rockett, K., Jallow, M., Pinder, M., Sisay-Joof, F. & Kwiatkowski, D. (2005) Investigation of malaria susceptibility determinants in the IFNG/IL26/IL22 genomic region. *Genes Immun* **6**: 312–318.

Kremsner, P.G., Grundmann, H., Neifer, S., Sliwa, K., Sahlmuller, G., Hegenscheid, B. & Bienzle, U. (1991) Pentoxifylline prevents murine cerebral malaria. *J Infect Dis* **164**: 605–608.

Kremsner, P.G., Winkler, S., Wildling, E., Prada, J., Bienzle, U., Graninger, W. & Nussler, A.K. (1996) High plasma levels of nitrogen oxides are associated with severe disease and correlate with rapid parasitological and clinical cure in *Plasmodium falciparum* malaria. *Trans R Soc Trop Med Hyg* **90**: 44–47.

Kumaratilake, L.M., Ferrante, A., Jaeger, T. & Morris-Jones, S.D. (1997) The role of complement, antibody, and tumor necrosis factor alpha in the killing of *Plasmodium falciparum* by the monocytic cell line THP-1. *Infect Immun* **65**: 5342–5345.

Kumaratilake, L.M., Ferrante, A. & Rzepczyk, C. (1991) The role of T lymphocytes in immunity to *Plasmodium falciparum*. Enhancement of neutrophil-mediated parasite killing by lymphotoxin and IFN-gamma: comparisons with tumor necrosis factor effects. *J Immunol* **146**: 762–767.

Kun, J.F., Mordmuller, B., Lell, B., Lehman, L.G., Luckner, D. & Kremsner, P.G. (1998) Polymorphism in promoter region of inducible nitric oxide synthase gene and protection against malaria. *Lancet* **351**: 265–266.

Kun, J.F., Klabunde, J., Lell, B., Luckner, D., Alpers, M., May, J., Meyer, C. & Kremsner, P.G. (1999) Association of the ICAM-1Kilifi mutation with protection against severe malaria in Lambarene, Gabon. *Am J Trop Med Hyg* **61**: 776–779.

Kun, J.F., Mordmuller, B., Perkins, D.J., May, J., Mercereau-Puijalon, O., Alpers, M., Weinberg, J.B. & Kremsner, P.G. (2001) Nitric oxide synthase 2(Lambarene)

(G-954C), increased nitric oxide production, and protection against malaria. *J Infect Dis* **184**: 330–336.

Kurtzhals, J.A., Adabayeri, V., Goka, B.Q., Akanmori, B.D., Oliver-Commey, J.O., Nkrumah, F.K., Behr, C. & Hviid, L. (1998) Low plasma concentrations of inter-leukin 10 in severe malarial anaemia compared with cerebral and uncomplicated malaria. *Lancet* **351**: 1768–1772.

Kurtzhals, J.A., Akanmori, B.D., Goka, B.Q., Adabayeri, V., Nkrumah, F.K., Behr, C. & Hviid, L. (1999) The cytokine balance in severe malarial anemia. *J Infect Dis* **180**: 1753–1755.

Kwiatkowski, D. (1995) Malarial toxins and the regulation of parasite density. *Parasitol Today* **11**: 206–212.

Kwiatkowski, D., Cannon, J.G., Manogue, K.R., Cerami, A., Dinarello, C.A. & Greenwood, B.M. (1989) Tumour necrosis factor production in *falciparum* malaria and its association with schizont rupture. *Clin Exp Immunol* **77**: 361–366.

Kwiatkowski, D., Hill, A.V., Sambou, I., Twumasi, P., Castracane, J., Manogue, K.R., Cerami, A., Brewster, D.R. & Greenwood, B.M. (1990) TNF concentration in fatal cerebral, non-fatal cerebral, and uncomplicated *Plasmodium falciparum* malaria. *Lancet* **336**: 1201–1204.

Kwiatkowski, D., Molyneux, M.E., Stephens, S., Curtis, N., Klein, N., Pointaire, P., Smit, M., Allan, R., Brewster, D.R., Grau, G.E., *et al.* (1993) Anti-TNF therapy inhibits fever in cerebral malaria. *Q J Med* **86**: 91–98.

Kwiatkowski, D.P. (2005) How malaria has affected the human genome and what human genetics can teach us about malaria. *Am J Hum Genet* **77**: 171–192.

Kyes, S., Horrocks, P. & Newbold, C. (2001) Antigenic variation at the infected red cell surface in malaria. *Annu Rev Microbiol* **55**: 673–707.

Lamas, S., Michel, T., Brenner, B.M. & Marsden, P.A. (1991) Nitric oxide synthesis in endothelial cells: evidence for a pathway inducible by TNF-alpha. *Am J Physiol* **261**: c634–c641.

Leiriao, P., Mota, M.M. & Rodriguez, A. (2005) Apoptotic Plasmodium-infected hepatocytes provide antigens to liver dendritic cells. *J Infect Dis* **191**: 1576–1581.

Leone, A.M., Palmer, R.M., Knowles, R.G., Francis, P.L., Ashton, D.S. & Moncada, S. (1991) Constitutive and inducible nitric oxide synthases incorporate molecular oxygen into both nitric oxide and citrulline. *J Biol Chem* **266**: 23790–23795.

Levesque, M.C., Hobbs, M.R., Anstey, N.M., Vaughn, T.N., Chancellor, J.A., Pole, A., Perkins, D.J., Misukonis, M.A., Chanock, S.J., Granger, D.L. & Weinberg, J.B. (1999) Nitric oxide synthase type 2 promoter polymorphisms, nitric oxide production, and disease severity in Tanzanian children with malaria. *J Infect Dis* **180**: 1994–2002.

Li, C., Corraliza, I. & Langhorne, J. (1999) A defect in interleukin-10 leads to enhanced malarial disease in *Plasmodium chabaudi chabaudi* infection in mice. *Infect Immun* **67**: 4435–4442.

Li, C., Sanni, L.A., Omer, F., Riley, E. & Langhorne, J. (2003) Pathology of *Plasmodium chabaudi chabaudi* infection and mortality in interleukin-10-deficient mice are ameliorated by anti-tumor necrosis factor alpha and exacerbated by anti-transforming growth factor beta antibodies. *Infect Immun* **71**: 4850–4856.

Looareesuwan, S., Wilairatana, P., Vannaphan, S., Wanaratana, V., Wenisch, C., Aikawa, M., Brittenham, G., Graninger, W. & Wernsdorfer, W.H. (1998) Pentoxifylline as an ancillary treatment for severe *falciparum* malaria in Thailand. *Am J Trop Med Hyg* **58**: 348–353.

Looareesuwan, S., Sjostrom, L., Krudsood, S., Wilairatana, P., Porter, R.S., Hills, F. & Warrell, D.A. (1999) Polyclonal anti-tumor necrosis factor-alpha Fab used as an ancillary treatment for severe malaria. *Am J Trop Med Hyg* **61**: 26–33.

Lopansri, B.K., Anstey, N.M., Weinberg, J.B., Stoddard, G.J., Hobbs, M.R., Levesque, M.C., Mwaikambo, E.D. & Granger, D.L. (2003) Low plasma arginine concentrations in children with cerebral malaria and decreased nitric oxide production. *Lancet* **361**: 676–678.

Lowell, C. (1997) Fundamentals of cell biology. In: Daniel P. Stites, A. I. T., Tristram G. Parslow (eds) *Medical Immunology*, 9 edn. Prentice-Hall International Inc, New Jersey.

Luoni, G., Verra, F., Arca, B., Sirima, B.S., Troye-Blomberg, M., Coluzzi, M., Kwiatkowski, D. & Modiano, D. (2001) Antimalarial antibody levels and IL4 polymorphism in the Fulani of West Africa. *Genes Immun* **2**: 411–414.

Luty, A.J., Lell, B., Schmidt-Ott, R., Lehman, L.G., Luckner, D., Greve, B., Matousek, P., Herbich, K., Schmid, D., Migot-Nabias, F., Deloron, P., Nussenzweig, R.S. & Kremsner, P.G. (1999) Interferon-gamma responses are associated with resistance to reinfection with *Plasmodium falciparum* in young African children. *J Infect Dis* **179**: 980–988.

Luty, A.J., Perkins, D.J., Lell, B., Schmidt-Ott, R., Lehman, L.G., Luckner, D., Greve, B., Matousek, P., Herbich, K., Schmid, D., Weinberg, J.B. & Kremsner, P.G. (2000) Low interleukin-12 activity in severe *Plasmodium falciparum* malaria. *Infect Immun* **68**: 3909–3915.

Luzzatto, L., Usanga, F.A. & Reddy, S. (1969) Glucose-6-phosphate dehydrogenase deficient red cells: resistance to infection by malarial parasites. *Science* **164**: 839–842.

Luzzatto, L., Nwachuku-Jarrett, E.S. & Reddy, S. (1970) Increased sickling of parasitized erythrocytes as mechanism of resistance against malaria in the sickle-cell trait. *Lancet* **1**: 319–321.

Luzzi, G.A., Merry, A.H., Newbold, C.I., Marsh, K. & Pasvol, G. (1991a) Protection by alpha-thalassaemia against *Plasmodium falciparum* malaria: modified surface antigen expression rather than impaired growth or cytoadherence. *Immunol Lett* **30**: 233–2340.

Luzzi, G.A., Merry, A.H., Newbold, C.I., Marsh, K., Pasvol, G. & Weatherall, D.J. (1991b) Surface antigen expression on *Plasmodium falciparum*-infected erythrocytes is modified in alpha- and beta-thalassemia. *J Exp Med* **173**: 785–791.

MacKinnon, M.J., Walker, P.R. & Rowe, J.A. (2002) *Plasmodium chabaudi*: rosetting in a rodent malaria model. *Exp Parasitol* **101**: 121–128.

Macpherson, G.G., Warrell, M.J., White, N.J., Looareesuwan, S. & Warrell, D.A. (1985) Human cerebral malaria. A quantitative ultrastructural analysis of parasitized erythrocyte sequestration. *Am J Pathol* **119**: 385–401.

Maier, A.G., Duraisingh, M.T., Reeder, J.C., Patel, S.S., Kazura, J.W., Zimmerman, P.A. & Cowman, A.F. (2003) *Plasmodium falciparum* erythrocyte invasion through glycophorin C and selection for Gerbich negativity in human populations. *Nat Med* **9**: 87–92.

Mannel, D.N. & Grau, G.E. (1997) Role of platelet adhesion in homeostasis and immunopathology. *Mol Pathol* **50:** 175–185.

May, J., Lell, B., Luty, A.J., Meyer, C.G. & Kremsner, P.G. (2000) Plasma interleukin-10:tumor necrosis factor (TNF)-alpha ratio is associated with TNF promoter variants and predicts malarial complications. *J Infect Dis* **182:** 1570–1573.

May, J., Lell, B., Luty, A.J., Meyer, C.G. & Kremsner, P.G. (2001) HLA-DQB1*0501-restricted Th1 type immune responses to *Plasmodium falciparum* liver stage antigen 1 protect against malaria anemia and reinfections. *J Infect Dis* **183:** 168–172.

Mayer, D.C., Kaneko, O., Hudson-Taylor, D.E., Reid, M.E. & Miller, L.H. (2001) Characterization of a *Plasmodium falciparum* erythrocyte-binding protein paralogous to EBA-175. *Proc Natl Acad Sci USA* **98:** 5222–5227.

Mazier, D., Nitcheu, J. & Idrissa-Boubou, M. (2000) Cerebral malaria and immunogenetics. *Parasite Immunol* **22:** 613–623.

McAdam, S.N., Boyson, J.E., Liu, X., Garber, T.L., Hughes, A.L., Bontrop, R.E. & Watkins, D.I. (1994) A uniquely high level of recombination at the HLA-B locus. *Proc Natl Acad Sci USA* **91:** 5893–5897.

McCormick, C.J., Craig, A., Roberts, D., Newbold, C.I. & Berendt, A.R. (1997) Intercellular adhesion molecule-1 and CD36 synergize to mediate adherence of *Plasmodium falciparum*-infected erythrocytes to cultured human microvascular endothelial cells. *J Clin Invest* **100:** 2521–2529.

McGilvray, I.D., Serghides, L., Kapus, A., Rotstein, O.D. & Kain, K.C. (2000) Nonopsonic monocyte/macrophage phagocytosis of *Plasmodium falciparum*-parasitized erythrocytes: a role for CD36 in malarial clearance. *Blood* **96:** 3231–3240.

McGuire, W., Hill, A.V., Allsopp, C.E., Greenwood, B.M. & Kwiatkowski, D. (1994) Variation in the TNF-alpha promoter region associated with susceptibility to cerebral malaria. *Nature* **371:** 508–510.

McGuire, W., Knight, J.C., Hill, A.V., Allsopp, C.E., Greenwood, B.M. & Kwiatkowski, D. (1999) Severe malarial anemia and cerebral malaria are associated with different tumor necrosis factor promoter alleles. *J Infect Dis* **179:** 287–290.

Meager, A. (1999) Cytokine regulation of cellular adhesion molecule expression in inflammation. *Cytokine Growth Factor Rev* **10:** 27–39.

Meding, S.J., Cheng, S.C., Simon-Haarhaus, B. & Langhorne, J. (1990) Role of gamma interferon during infection with *Plasmodium chabaudi chabaudi*. *Infect Immun* **58:** 3671–3678.

Mellouk, S., Green, S.J., Nacy, C.A. & Hoffman, S.L. (1991) IFN-gamma inhibits development of *Plasmodium berghei* exoerythrocytic stages in hepatocytes by an L-arginine-dependent effector mechanism. *J Immunol* **146:** 3971–3976.

Meyer, C.G., May, J., Luty, A.J., Lell, B. & Kremsner, P.G. (2002) TNFalpha-308A associated with shorter intervals of *Plasmodium falciparum* reinfections. *Tissue Antigens* **59:** 287–292.

Migot-Nabias, F., Luty, A.J., Minh, T.N., Fajardy, I., Tamouza, R., Marzais, F., Charron, D., Danze, P.M., Renaut, A. & Deloron, P. (2001) HLA alleles in relation to specific immunity to liver stage antigen-1 from *Plasmodium falciparum* in Gabon. *Genes Immun* **2:** 4–10.

Miller, L.H., Mason, S.J., Clyde, D.F. & McGinniss, M.H. (1976) The resistance factor to *Plasmodium vivax* in blacks. The Duffy-blood-group genotype, fyfy. *N Engl J Med* 295: 302–304.

Miller, L.H., Good, M.F. & Milon, G. (1994) Malaria pathogenesis. *Science* 264: 1878–1883.

Miller, L.H., Baruch, D.I., Marsh, K. & Doumbo, O.K. (2002) The pathogenic basis of malaria. *Nature* 415: 673–679.

Min-Oo, G. & Gros, P. (2005) Erythrocyte variants and the nature of their malaria protective effect. *Cell Microbiol* 7: 753–763.

Min-Oo, G., Fortin, A., Tam, M.F., Nantel, A., Stevenson, M.M. & Gros, P. (2003) Pyruvate kinase deficiency in mice protects against malaria. *Nat Genet* 35: 357–362.

Min-Oo, G., Fortin, A., Tam, M.F., Gros, P. & Stevenson, M.M. (2004) Phenotypic expression of pyruvate kinase deficiency and protection against malaria in a mouse model. *Genes Immun* 5: 168–175.

Mitchell, G.H., Hadley, T.J., McGinniss, M.H., Klotz, F.W. & Miller, L.H. (1986) Invasion of erythrocytes by *Plasmodium falciparum* malaria parasites: evidence for receptor heterogeneity and two receptors. *Blood* 67: 1519–1521.

Modiano, D., Luoni, G., Sirima, B.S., Simpore, J., Verra, F., Konate, A., Rastrelli, E., Olivieri, A., Calissano, C., Paganotti, G.M., D'Urbano, L., Sanou, I., Swadogo, A., Modiano, G. & Coluzzi, M. (2001) Haemoglobin C protects against clinical *Plasmodium falciparum* malaria. *Nature* 414: 305–308.

Mohan, K., Moulin, P. & Stevenson, M.M. (1997) Natural killer cell cytokine production, not cytotoxicity, contributes to resistance against blood-stage *Plasmodium chabaudi* as infection. *J Immunol* 159: 4990–4998.

Mombo, L.E., Ntoumi, F., Bisseye, C., Ossari, S., Lu, C.Y., Nagel, R.L. & Krishnamoorthy, R. (2003) Human genetic polymorphisms and asymptomatic *Plasmodium falciparum* malaria in Gabonese schoolchildren. *Am J Trop Med Hyg* 68: 186–190.

Mota, M.M. & Rodriguez, A. (2004) Migration through host cells: the first steps of *Plasmodium* sporozoites in the mammalian host. *Cell Microbiol* 6: 1113–1118.

Mota, M.M., Jarra, W., Hirst, E., Patnaik, P.K. & Holder, A.A. (2000) *Plasmodium chabaudi*-infected erythrocytes adhere to CD36 and bind to microvascular endothelial cells in an organ-specific way. *Infect Immun* 68: 4135–4144.

Mota, M.M., Pradel, G., Vanderberg, J.P., Hafalla, J.C., Frevert, U., Nussenzweig, R.S., Nussenzweig, V. & Rodriguez, A. (2001) Migration of *Plasmodium* sporozoites through cells before infection. *Science* 291: 141–144.

Moulds, J.M., Nickells, M.W., Moulds, J.J., Brown, M.C. & Atkinson, J.P. (1991) The C3b/C4b receptor is recognized by the Knops, McCoy, Swain-Langley, and York blood group antisera. *J Exp Med* 173: 1159–1163.

Moulds, J.M., Kassambara, L., Middleton, J.J., Baby, M., Sagara, I., Guindo, A., Coulibaly, S., Yalcouye, D., Diallo, D.A., Miller, L. & Doumbo, O. (2000) Identification of complement receptor one (CR1) polymorphisms in West Africa. *Genes Immun* 1: 325–329.

Moulds, J.M., Zimmerman, P.A., Doumbo, O.K., Kassambara, L., Sagara, I., Diallo, D.A., Atkinson, J.P., Krych-Goldberg, M., Hauhart, R.E., Hourcade, D.E., McNamara, D.T., Birmingham, D.J., Rowe, J.A., Moulds, J.J. & Miller, L.H.

(2001) Molecular identification of Knops blood group polymorphisms found in long homologous region D of complement receptor 1. *Blood* **97**: 2879–2885.

Nagamine, Y., Hayano, M., Kashiwamura, S., Okamura, H., Nakanishi, K., Krudsod, S., Wilairatana, P., Looareesuwan, S. & Kojima, S. (2003) Involvement of interleukin-18 in severe *Plasmodium falciparum* malaria. *Trans R Soc Trop Med Hyg* **97**: 236–241.

Nagayasu, E., Ito, M., Akaki, M., Nakano, Y., Kimura, M., Looareesuwan, S. & Aikawa, M. (2001) CR1 density polymorphism on erythrocytes of *falciparum* malaria patients in Thailand. *Am J Trop Med Hyg* **64**: 1–5.

Neote, K., Mak, J.Y., Kolakowski, L.F., Jr. & Schall, T.J. (1994) Functional and biochemical analysis of the cloned Duffy antigen: identity with the red blood cell chemokine receptor. *Blood* **84**: 44–52.

Newman, P.J., Berndt, M.C., Gorski, J., White, G.C., 2nd, Lyman, S., Paddock, C. & Muller, W.A. (1990) PECAM-1 (CD31) cloning and relation to adhesion molecules of the immunoglobulin gene superfamily. *Science* **247**: 1219–1222.

Nussenblatt, V., Mukasa, G., Metzger, A, Ndeezi, G., Garrett, E. & Semba, R.D. (2001) Anemia and interleukin-10, tumor necrosis factor alpha, and erythropoietin levels among children with acute, uncomplicated *Plasmodium falciparum* malaria. *Clin Diagn Lab Immunol* **8**: 1164–1170.

Nussenzweig, R.S., Vanderberg, J., Most, H. & Orton, C. (1967) Protective immunity produced by the injection of X-irradiated sporozoites of *Plasmodium berghei*. *Nature* **216**: 160–162.

Nussler, A., Drapier, J.C., Renia, L., Pied, S., Miltgen, F., Gentilini, M. & Mazier, D. (1991) L-arginine-dependent destruction of intrahepatic malaria parasites in response to tumor necrosis factor and/or interleukin 6 stimulation. *Eur J Immunol* **21**: 227–230.

Ockenhouse, C.F., Schulman, S. & Shear, H.L. (1984) Induction of crisis forms in the human malaria parasite *Plasmodium falciparum* by gamma-interferon-activated, monocyte-derived macrophages. *J Immunol* **133**: 1601–1608.

Ockenhouse, C.F., Tegoshi, T., Maeno, Y., Benjamin, C., Ho, M., Kan, K.E., Thway, Y., Win, K., Aikawa, M. & Lobb, R.R. (1992) Human vascular endothelial cell adhesion receptors for *Plasmodium falciparum*-infected erythrocytes: roles for endothelial leukocyte adhesion molecule 1 and vascular cell adhesion molecule 1. *J Exp Med* **176**: 1183–1189.

Odeh, M. (2001) The role of tumour necrosis factor-alpha in the pathogenesis of complicated *falciparum* malaria. *Cytokine* **14**: 11–18.

Ohashi, J., Naka, I., Patarapotikul, J., Hananantachai, H., Looareesuwan, S. & Tokunaga, K. (2001) Absence of association between the allele coding methionine at position 29 in the N-terminal domain of ICAM-1 (ICAM-1(Kilifi)) and severe malaria in the northwest of Thailand. *Jpn J Infect Dis* **54**: 114–116.

Ohashi, J., Naka, I., Patarapotikul, J., Hananantachai, H., Looareesuwan, S. & Tokunaga, K. (2002) Significant association of longer forms of CCTTT microsatellite repeat in the inducible nitric oxide synthase promoter with severe malaria in Thailand. *J Infect Dis* **186**: 578–581.

Omi, K., Ohashi, J., Patarapotikul, J., Hananantachai, H., Naka, I., Looareesuwan, S. & Tokunaga, K. (2002) Fcgamma receptor IIa and IIIb polymorphisms are associated with susceptibility to cerebral malaria. *Parasitol Int* **51**: 361–366.

Omi, K., Ohashi, J., Patarapotikul, J., Hananantachai, H., Naka, I., Looareesuwan, S. & Tokunaga, K. (2003) CD36 polymorphism is associated with protection from cerebral malaria. *Am J Hum Genet* **72**: 364–374.

Orago, A.S. & Facer, C.A. (1991) Cytotoxicity of human natural killer (NK) cell subsets for *Plasmodium falciparum* erythrocytic schizonts: stimulation by cytokines and inhibition by neomycin. *Clin Exp Immunol* **86**: 22–29.

Oswald, I.P., Eltoum, I., Wynn, T.A., Schwartz, B., Caspar, P., Paulin, D., Sher, A. & James, S.L. (1994) Endothelial cells are activated by cytokine treatment to kill an intravascular parasite, *Schistosoma mansoni*, through the production of nitric oxide. *Proc Natl Acad Sci USA* **91**: 999–1003.

Pain, A., Ferguson, D.J., Kai, O., Urban, B.C., Lowe, B., Marsh, K. & Roberts, D.J. (2001a) Platelet-mediated clumping of *Plasmodium falciparum*-infected erythrocytes is a common adhesive phenotype and is associated with severe malaria. *Proc Natl Acad Sci USA* **98**: 1805–1810.

Pain, A., Urban, B.C., Kai, O., Casals-Pascual, C., Shafi, J., Marsh, K. & Roberts, D.J. (2001b) A non-sense mutation in CD36 gene is associated with protection from severe malaria. *Lancet* **357**: 1502–1503.

Parikh, S., Dorsey, G. & Rosenthal, P.J. (2004) Host polymorphisms and the incidence of malaria in Ugandan children. *Am J Trop Med Hyg* **71**: 750–753.

Pasvol, G., Weatherall, D.J. & Wilson, R.J. (1978) Cellular mechanism for the protective effect of haemoglobin S against *P. falciparum* malaria. *Nature* **274**: 701–703.

Pasvol, G., Wainscoat, J.S. & Weatherall, D.J. (1982) Erythrocytes deficiency in glycophorin resist invasion by the malarial parasite *Plasmodium falciparum*. *Nature* **297**: 64–66.

Patel, S.S., Mehlotra, R.K., Kastens, W., Mgone, C.S., Kazura, J.W. & Zimmerman, P.A. (2001) The association of the glycophorin C exon 3 deletion with ovalocytosis and malaria susceptibility in the Wosera, Papua New Guinea. *Blood* **98**: 3489–3491.

Perkins, D.J., Kremsner, P.G., Schmid, D., Misukonis, M.A., Kelly, M.A. & Weinberg, J.B. (1999) Blood mononuclear cell nitric oxide production and plasma cytokine levels in healthy Gabonese children with prior mild or severe malaria. *Infect Immun* **67**: 4977–4981.

Pfefferkorn, E.R. & Guyre, P.M. (1984) Inhibition of growth of *Toxoplasma gondii* in cultured fibroblasts by human recombinant gamma interferon. *Infect Immun* **44**: 211–216.

Pichyangkul, S., Saengkrai, P. & Webster, H.K. (1994) *Plasmodium falciparum* pigment induces monocytes to release high levels of tumor necrosis factor-alpha and interleukin-1 beta. *Am J Trop Med Hyg* **51**: 430–435.

Piguet, P.F., Kan, C.D., Vesin, C., Rochat, A., Donati, Y. & Barazzone, C. (2001) Role of CD40-CVD40l in mouse severe malaria. *Am J Pathol* **159**: 733–742.

Pober, I.S. & Cotran, R.S. (1990) Cytokines and endothelial cell biology. *Physiol Rev* **70**: 427–451.

Pober, J.S., Gimbrone, M.A., Jr., Lapierre, L.A., Mendrick, D.L., Fiers, W., Rothlein, R. & Springer, T.A. (1986) Overlapping patterns of activation of human endothelial cells by interleukin 1, tumor necrosis factor, and immune interferon. *J Immunol* **137**: 1893–1896.

84   CH. 3.   THE RELEVANCE OF HOST GENES IN MALARIA

Praba-Egge, A.D., Montenegro, S., Cogswell, F.B., Hopper, T. & James, M.A. (2002) Cytokine responses during acute simian *Plasmodium cynomolgi* and *Plasmodium knowlesi* infections. *Am J Trop Med Hyg* 67: 586–596.

Prudhomme, J.G., Sherman, I.W., Land, K.M., Moses, A.V., Stenglein, S. & Nelson, J.A. (1996) Studies of *Plasmodium falciparum* cytoadherence using immortalized human brain capillary endothelial cells. *Int J Parasitol* 26: 647–655.

Rabellino, E.M., Ross, G.D. & Polley, M.J. (1978) Membrane receptors of mouse leukocytes. I. Two types of complement receptors for different regions of C3. *J Immunol* 120: 879–885.

Rasti, N., Wahlgren, M. & Chen, Q. (2004) Molecular aspects of malaria pathogenesis. *Fems Immunol Med Microbiol* 41: 9–26.

Ravetch, J.V. & Kinet, J.P. (1991) Fc receptors. *Annu Rev Immunol* 9: 457–492.

Rayner, J.C., Vargas-Serrato, E., Huber, C.S., Galinski, M.R. & Barnwell, J.W. (2001) A *Plasmodium falciparum* homologue of *Plasmodium vivax* reticulocyte binding protein (PvRBP1) defines a trypsin-resistant erythrocyte invasion pathway. *J Exp Med* 194: 1571–1581.

Reed, M.B., Caruana, S.R., Batchelor, A.H., Thompson, J.K., Crabb, B.S. & Cowman, A.F. (2000) Targeted disruption of an erythrocyte binding antigen in *Plasmodium falciparum* is associated with a switch toward a sialic acid-independent pathway of invasion. *Proc Natl Acad Sci USA* 97: 7509–7514.

Richards, A.L. (1997) Tumour necrosis factor and associated cytokines in the host's response to malaria. *Int J Parasitol* 27: 1251–1263.

Richer, J. & Chudley, A.E. (2005) The hemoglobinopathies and malaria. *Clin Genet* 68: 332–336.

Rieckmann, K.H., Carson, P.E., Beaudoin, R.L., Cassells, J.S. & Sell, K.W. (1974) Letter: sporozoite induced immunity in man against an Ethiopian strain of *Plasmodium falciparum*. *Trans R Soc Trop Med Hyg* 68: 258–259.

Riley, E.M., Jakobsen, P.H., Allen, S.J., Wheeler, J.G., Bennett, S., Jepsen, S. & Greenwood, B.M. (1991) Immune response to soluble exoantigens of *Plasmodium falciparum* may contribute to both pathogenesis and protection in clinical malaria: evidence from a longitudinal, prospective study of semi-immune African children. *Eur J Immunol* 21: 1019–1025.

Ringwald, P., Peyron, F., Vuillez, J.P., Touze, J.E., Le Bras, J. & Deloron, P. (1991) Levels of cytokines in plasma during *Plasmodium falciparum* malaria attacks. *J Clin Microbiol* 29: 2076–2078.

Roberts, D.D., Sherwood, J.A., Spitalnik, S.L., Panton, L.J., Howard, R.J., Dixit, V.M., Frazier, W.A., Miller, L.H. & Ginsburg, V. (1985) Thrombospondin binds *falciparum* malaria parasitized erythrocytes and may mediate cytoadherence. *Nature* 318: 64–66.

Roberts, D.J., Craig, A.G., Berendt, A.R., Pinches, R., Nash, G., Marsh, K. & Newbold, C.I. (1992) Rapid switching to multiple antigenic and adhesive phenotypes in malaria. *Nature* 357: 689–692.

Rockett, K.A., Targett, G.A. & Playfair, J.H. (1988) Killing of blood-stage *Plasmodium falciparum* by lipid peroxides from tumor necrosis serum. *Infect Immun* 56: 3180–3183.

Rockett, K.A., Awburn, M.M., Aggarwal, B.B., Cowden, W.B. & Clark, I.A. (1992) *In vivo* induction of nitrite and nitrate by tumor necrosis factor, lymphotoxin, and interleukin-1: possible roles in malaria. *Infect Immun* 60: 3725–3730.

Rockett, K.A., Awburn, M.M., Rockett, E.J., Cowden, W.B. & Clark, I.A. (1994) Possible role of nitric oxide in malarial immunosuppression. *Parasite Immunol* **16**: 243–249.

Rogerson, S.J. & Beeson, J.G. (1999) The placenta in malaria: mechanisms of infection, disease and foetal morbidity. *Ann Trop Med Parasitol* **93** suppl 1, s35–s42.

Rogerson, S.J., Chaiyaroj, S.C., Ng, K., Reeder, J.C. & Brown, G.V. (1995) Chondroitin sulfate a is a cell surface receptor for *Plasmodium falciparum*-infected erythrocytes. *J Exp Med* **182**: 15–20.

Romagnani, S. (1995) Biology of human TH1 and TH2 cells. *J Clin Immunol* **15**: 121–9.

Rossi-Bergmann, B., Muller, I. & Godinho, E.B. (1993) TH1 and TH2 T-cell subsets are differentially activated by macrophages and B cells in murine leishmaniasis. *Infect Immun* **61**: 2266–2269.

Rothstein, A., Cabantchik, Z.I. & Knauf, P. (1976) Mechanism of anion transport in red blood cells: role of membrane proteins. *Fed Proc* **35**: 3–10.

Rotman, H.L., Daly, T.M., Clynes, R. & Long, C.A. (1998) Fc receptors are not required for antibody-mediated protection against lethal malaria challenge in a mouse model. *J Immunol* **161**: 1908–1912.

Rowe, J.A., Moulds, J.M., Newbold, C.I. & Miller, L.H. (1997) *P. Falciparum* rosetting mediated by a parasite-variant erythrocyte membrane protein and complement-receptor 1. *Nature* **388**: 292–295.

Rowe, J.A., Rogerson, S.J., Raza, A., Moulds, J.M., Kazatchkine, M.D., Marsh, K., Newbold, C.I., Atkinson, J.P. & Miller, L.H. (2000) Mapping of the region of complement receptor (CR) 1 required for *Plasmodium falciparum* rosetting and demonstration of the importance of CR1 in rosetting in field isolates. *J Immunol* **165**: 6341–6346.

Rowe, J.A., Raza, A., Diallo, D.A., Baby, M., Poudiougo, B., Coulibaly, D., Cockburn, I.A., Middleton, J., Lyke, K.E., Plowe, C.V., Doumbo, O.K. & Moulds, J.M. (2002) Erythrocyte CR1 expression level does not correlate with a HindIII restriction fragment length polymorphism in Africans; implications for studies on malaria susceptibility. *Genes Immun* **3**: 497–500.

Rudin, W., Favre, N., Bordmann, G. & Ryffel, B. (1997) Interferon-gamma is essential for the development of cerebral malaria. *Eur J Immunol* **27**: 810–815.

Sabeti, P., Usen, S., Farhadian, S., Jallow, M., Doherty, T., Newport, M., Pinder, M., Ward, R. & Kwiatkowski, D. (2002) CD40l association with protection from severe malaria. *Genes Immun* **3**: 286–291.

Saeftel, M., Krueger, A., Arriens, S., Heussler, V., Racz, P., Fleischer, B., Brombacher, F. & Hoerauf, A. (2004) Mice deficient in interleukin-4 (IL-4) or IL-4 receptor alpha have higher resistance to sporozoite infection with *Plasmodium berghei* (ANKA) than do naive wild-type mice. *Infect Immun* **72**: 322–331.

Sanni, L.A., Jarra, W., Li, C. & Langhorne, J. (2004) Cerebral edema and cerebral hemorrhages in interleukin-10-deficient mice infected with *Plasmodium chabaudi*. *Infect Immun* **72**: 3054–3058.

Sartelet, H., Garraud, O., Rogier, C., Milko-Sartelet, I., Kaboret, Y., Michel, G., Roussilhon, C., Huerre, M. & Gaillard, D. (2000) Hyperexpression of ICAM-1 and CD36 in placentas infected with *Plasmodium falciparum*: a possible role of these molecules in sequestration of infected red blood cells in placentas. *Histopathology* **36**: 62–68.

Schofield, A.E., Reardon, D.M. & Tanner, M.J. (1992) Defective anion transport activity of the abnormal band 3 in hereditary ovalocytic red blood cells. *Nature* **355**: 836–838.

Schwarzer, E., Kuhn, H., Valente, E. & Arese, P. (2003) Malaria-parasitized erythrocytes and hemozoin nonenzymatically generate large amounts of hydroxy fatty acids that inhibit monocyte functions. *Blood* **101**: 722–728.

Seixas, E., Fonseca, L. & Langhorne, J. (2002) The influence of gammadelta T cells on the CD4+ T cell and antibody response during a primary *Plasmodium chabaudi chabaudi* infection in mice. *Parasite Immunol* **24**: 131–140.

Serghides, L., Crandall, I., Hull, E. & Kain, K.C. (1998) The *Plasmodium falciparum*-CD36 interaction is modified by a single amino acid substitution in CD36. *Blood* **92**: 1814–1819.

Serghides, L., Smith, T.G., Patel, S.N. & Kain, K.C. (2003) CD36 and malaria: friends or foes? *Trends Parasitol* **19**: 461–469.

Sharief, M.K. & Thompson, E.J. (1992) *In vivo* relationship of tumor necrosis factor-alpha to blood-brain barrier damage in patients with active multiple sclerosis. *J Neuroimmunol* **38**: 27–33.

Sharief, M.K., Ciardi, M. & Thompson, E.J. (1992) Blood-brain barrier damage in patients with bacterial meningitis: association with tumor necrosis factor-alpha but not interleukin-1 beta. *J Infect Dis* **166**: 350–358.

Shear, H.L., Srinivasan, R., Nolan, T. & Ng, C. (1989) Role of IFN-gamma in lethal and nonlethal malaria in susceptible and resistant murine hosts. *J Immunol* **143**: 2038–2044.

Shear, H.L., Roth, E.F., Jr., Fabry, M.E., Costantini, F.D., Pachnis, A., Hood, A. & Nagel, R.L. (1993) Transgenic mice expressing human sickle hemoglobin are partially resistant to rodent malaria. *Blood* **81**: 222–226.

Sherman, I.W., Eda, S. & Winograd, E. (2003) Cytoadherence and sequestration in *Plasmodium falciparum*: defining the ties that bind. *Microbes Infect* **5**: 897–909.

Shimizu, H., Tamam, M., Soemantri, A. & Ishida, T. (2005) Glucose-6-phosphate dehydrogenase deficiency and southeast Asian ovalocytosis in asymptomatic *Plasmodium* carriers in Sumba Island, Indonesia. *J Hum Genet* **50**: 420–424.

Shiu, Y.T., Udden, M.M. & McIntire, L.V. (2000) Perfusion with sickle erythrocytes up-regulates ICAM-1 and VCAM-1 gene expression in cultured human endothelial cells. *Blood* **95**: 3232–3241.

Silamut, K., Phu, N.H., Whitty, C., Turner, G.D., Louwrier, K., Mai, N.T., Simpson, J.A., Hien, T.T. & White, N.J. (1999) A quantitative analysis of the microvascular sequestration of malaria parasites in the human brain. *Am J Pathol* **155**: 395–410.

Sobolewski, P., Gramaglia, I., Frangos, J., Intaglietta, M. & van der Heyde, H.C. (2005a) Nitric oxide bioavailability in malaria. *Trends Parasitol* **21**: 415–422.

Sobolewski, P., Gramaglia, I., Frangos, J.A., Intaglietta, M. & van der Heyde, H. (2005b) *Plasmodium berghei* resists killing by reactive oxygen species. *Infect Immun* **73**: 6704–6710.

Solomon, J.B., Forbes, M.G. & Solomon, G.R. (1985) A possible role for natural killer cells in providing protection against *Plasmodium berghei* in early stages of infection. *Immunol Lett* **9**: 349–352.

Stach, J.L., Dufrenoy, E., Roffi, J. & Bach, M.A. (1986) T-cell subsets and natural killer activity in *Plasmodium falciparum*-infected children. *Clin Immunol Immunopathol* **38**: 129–134.

Steers, N., Schwenk, R., Bacon, D.J., Berenzon, D., Williams, J. & Krzych, U. (2005) The immune status of kupffer cells profoundly influences their responses to infectious *Plasmodium berghei* sporozoites. *Eur J Immunol* **35**: 2335–2346.

Stevenson, M.M. & Riley, E.M. (2004) Innate immunity to malaria. *Nat Rev Immunol* **4**: 169–180.

Stevenson, M.M., Tam, M.F., Belosevic, M., van der Meide, P.H. & Podoba, J.E. (1990a) Role of endogenous gamma interferon in host response to infection with blood-stage *Plasmodium chabaudi* as. *Infect Immun* **58**: 3225–3232.

Stevenson, M.M., Tam, M.F. & Nowotarski, M. (1990b) Role of interferon-gamma and tumor necrosis factor in host resistance to *Plasmodium chabaudi* as. *Immunol Lett* **25**: 115–121.

Stevenson, M.M., Tam, M.F., Wolf, S.F. & Sher, A. (1995) IL-12-induced protection against blood-stage *Plasmodium chabaudi* as requires IFN-gamma and TNF-alpha and occurs via a nitric oxide-dependent mechanism. *J Immunol* **155**: 2545–2556.

Stirnadel, H.A., Stockle, M., Felger, I., Smith, T., Tanner, M. & Beck, H.P. (1999) Malaria infection and morbidity in infants in relation to genetic polymorphisms in Tanzania. *Trop Med Int Health* **4**: 187–193.

Stoltenburg-Didinger, G., Neifer, S., Bienzle, U., Eling, W.M. & Kremsner, P.G. (1993) Selective damage of hippocampal neurons in murine cerebral malaria prevented by pentoxifylline. *J Neurol Sci* **114**: 20–24.

Strieter, R.M., Remick, D.G., Ward, P.A., Spengler, R.N., Lynch, J.P., 3rd, Larrick, J. & Kunkel, S.L. (1988) Cellular and molecular regulation of tumor necrosis factor-alpha production by pentoxifylline. *Biochem Biophys Res Commun* **155**: 1230–1236.

Stubbs, J., Simpson, K.M., Triglia, T., Plouffe, D., Tonkin, C.J., Duraisingh, M.T., Maier, A.G., Winzeler, E.A. & Cowman, A.F. (2005) Molecular mechanism for switching of *P. falciparum* invasion pathways into human erythrocytes. *Science* **309**: 1384–1387.

Sugiyama, T., Cuevas, L.E., Bailey, W., Makunde, R., Kawamura, K., Kobayashi, M., Masuda, H. & Hommel, M. (2001) Expression of intercellular adhesion molecule 1 (ICAM-1) in *Plasmodium falciparum*-infected placenta. *Placenta* **22**: 573–579.

Sullivan, G.W., Carper, H.T., Novick, W.J., Jr. & Mandell, G.L. (1988) Inhibition of the inflammatory action of interleukin-1 and tumor necrosis factor (alpha) on neutrophil function by pentoxifylline. *Infect Immun* **56**: 1722–1729.

Sun, G., Chang, W.L., Li, J., Berney, S.M., Kimpel, D. & van der Heyde, H.C. (2003) Inhibition of platelet adherence to brain microvasculature protects against severe *Plasmodium berghei* malaria. *Infect Immun* **71**: 6553–6561.

Suzuki, Y., Orellana, M.A., Schreiber, R.D. & Remington, J.S. (1988) Interferon-gamma: the major mediator of resistance against *Toxoplasma gondii*. *Science* **240**: 516–518.

Tachado, S.D., Gerold, P., McConville, M.J., Baldwin, T., Quilici, D., Schwarz, R.T. & Schofield, L. (1996) Glycosylphosphatidylinositol toxin of *Plasmodium* induces nitric oxide synthase expression in macrophages and vascular endothelial cells by a protein tyrosine kinase-dependent and protein kinase C-dependent signaling pathway. *J Immunol* **156:** 1897–1907.

Taverne, J., Tavernier, J., Fiers, W. & Playfair, J.H. (1987) Recombinant tumour necrosis factor inhibits malaria parasites *in vivo* but not *in vitro*. *Clin Exp Immunol* **67:** 1–4.

Taverne, J., Bate, C.A., Kwiatkowski, D., Jakobsen, P.H. & Playfair, J.H. (1990a) Two soluble antigens of *Plasmodium falciparum* induce tumor necrosis factor release from macrophages. *Infect Immun* **58:** 2923–2928.

Taverne, J., Bate, C.A., Sarkar, D.A., Meager, A., Rook, G.A. & Playfair, J.H. (1990b) Human and murine macrophages produce TNF in response to soluble antigens of *Plasmodium falciparum*. *Parasite Immunol* **12:** 33–43.

Taylor-Robinson, A.W. (1995) Regulation of immunity to malaria: valuable lessons learned from murine models. *Parasitol Today* **11:** 334–342.

Taylor-Robinson, A.W., Phillips, R.S., Severn, A., Moncada, S. & Liew, F.Y. (1993) The role of TH1 and TH2 cells in a rodent malaria infection. *Science* **260:** 1931–1934.

Taylor, T.E., Fu, W.J., Carr, R.A., Whitten, R.O., Mueller, J.S., Fosiko, N.G., Lewallen, S., Liomba, N.G. & Molyneux, M.E. (2004) Differentiating the pathologies of cerebral malaria by postmortem parasite counts. *Nat Med* **10:** 143–145.

Todryk, S.M. & Walther, M. (2005) Building better T-cell-inducing malaria vaccines. *Immunology* **115:** 163–169.

Tokumasu, F., Fairhurst, R.M., Ostera, G.R., Brittain, N.J., Hwang, J., Wellems, T.E. & Dvorak, A. (2005) Band 3 modifications in *Plasmodium falciparum*-infected AA and CC erythrocytes assayed by autocorrelation analysis using quantum dots. *J Cell Sci* **118:**, 1091–1098.

Torre, D., Giola, M., Speranza, F., Matteelli, A., Basilico, C. & Biondi, G. (2001) Serum levels of interleukin-18 in patients with uncomplicated *Plasmodium falciparum* malaria. *Eur Cytokine Netw* **12:** 361–364.

Torre, D., Speranza, F., Giola, M., Matteelli, A., Tambini, R. & Biondi, G. (2002) Role of TH1 and TH2 cytokines in immune response to uncomplicated *Plasmodium falciparum* malaria. *Clin Diagn Lab Immunol* **9:** 348–351.

Tournamille, C., Colin, Y., Cartron, J.P. & Le Van Kim, C. (1995) Disruption of a GATA motif in the Duffy gene promoter abolishes erythroid gene expression in Duffy-negative individuals. *Nat Genet* **10:** 224–228.

Traore, B., Muanza, K., Looareesuwan, S., Supavej, S., Khusmith, S., Danis, M., Viriyavejakul, P. & Gay, F. (2000) Cytoadherence characteristics of *Plasmodium falciparum* isolates in Thailand using an *in vitro* human lung endothelial cells model. *Am J Trop Med Hyg* **62:** 38–44.

Treutiger, C.J., Heddini, A., Fernandez, V., Muller, W.A. & Wahlgren, M. (1997) PECAM-1/CD31, an endothelial receptor for binding *Plasmodium falciparum*-infected erythrocytes. *Nat Med* **3:** 1405–1408.

Troye-Blomberg, M., Riley, E.M., Kabilan, L., Holmberg, M., Perlmann, H., Andersson, U., Heusser, C.H. & Perlmann, P. (1990) Production by activated human T cells of interleukin 4 but not interferon-gamma is associated with elevated levels of serum antibodies to activating malaria antigens. *Proc Natl Acad Sci USA* **87:** 5484–5488.

Turner, G.D., Morrison, H., Jones, M., Davis, T.M., Looareesuwan, S., Buley, I.D., Gatter, K.C., Newbold, C.I., Pukritayakamee, S., Nagachinta, B., *et al.* (1994)

An immunohistochemical study of the pathology of fatal malaria. Evidence for widespread endothelial activation and a potential role for intercellular adhesion molecule-1 in cerebral sequestration. *Am J Pathol* **145**: 1057–1069.

Ubalee, R., Suzuki, F., Kikuchi, M., Tasanor, O., Wattanagoon, Y., Ruangweerayut, R., Na-Bangchang, K, Karbwang, J., Kimura, A., Itoh, K., Kanda, T. & Hirayama, K. (2001) Strong association of a tumor necrosis factor-alpha promoter allele with cerebral malaria in Myanmar. *Tissue Antigens* **58**: 407–410.

Udomsangpetch, R., Wahlin, B., Carlson, J., Berzins, K., Torii, M., Aikawa, M., Perlmann, P. & Wahlgren, M. (1989) *Plasmodium falciparum*-infected erythrocytes form spontaneous erythrocyte rosettes. *J Exp Med* **169**: 1835–1840.

Udomsangpetch, R., Webster, H.K., Pattanapanyasat, K., Pitchayangkul, S. & Thaithong, S. (1992) Cytoadherence characteristics of rosette-forming *Plasmodium falciparum*. *Infect Immun* **60**: 4483–4490.

Urban, B.C., Ferguson, D.J., Pain, A., Willcox, N., Plebanski, M., Austyn, J.M. & Roberts, D.J. (1999) *Plasmodium falciparum*-infected erythrocytes modulate the maturation of dendritic cells. *Nature* **400**: 73–77.

Urban, B.C., Willcox, N. & Roberts, D.J. (2001) A role for CD36 in the regulation of dendritic cell function. *Proc Natl Acad Sci USA* **98**: 8750–8755.

Van den Broek, M.F., Muller, U., Huang, S., Zinkernagel, R.M. & Aguet, M. (1995) Immune defence in mice lacking type I and/or type II interferon receptors. *Immunol Rev* **148**: 5–18.

Van der Heyde, H.C., Pepper, B., Batchelder, J., Cigel, F. & Weidanz, W.P. (1997) The time course of selected malarial infections in cytokine-deficient mice. *Exp Parasitol* **85**: 206–213.

Van der Poll, T., Jansen, J., Levi, M., Ten Cate, H., Ten Cate, J.W. & van Deventer, S.J. (1994) Regulation of interleukin 10 release by tumor necrosis factor in humans and chimpanzees. *J Exp Med* **180**: 1985–1988.

Van Hensbroek, M.B., Palmer, A., Onyiorah, E., Schneider, G., Jaffar, S., Dolan, G., Memming, H., Frenkel, J., Enwere, G., Bennett, S., Kwiatkowski, D. & Greenwood, B. (1996) The effect of a monoclonal antibody to tumor necrosis factor on survival from childhood cerebral malaria. *J Infect Dis* **174**: 1091–1097.

Verra, F., Luoni, G., Calissano, C., Troye-Blomberg, M., Perlmann, P., Perlmann, H., Arca, B., Sirima, B.S., Konate, A., Coluzzi, M., Kwiatkowski, D. & Modiano, D. (2004) IL4-589C/T polymorphism and IgE levels in severe malaria. *Acta Trop* **90**: 205–209.

Vogt, A.M., Barragan, A., Chen, Q., Kironde, F., Spillmann, D. & Wahlgren, M. (2003) Heparan sulfate on endothelial cells mediates the binding of *Plasmodium falciparum*-infected erythrocytes via the DBL1alpha domain of PfEMP1. *Blood* **101**: 2405–2411.

Von der Weid, T., Kopf, M., Kohler, G. & Langhorne, J. (1994) The immune response to *Plasmodium chabaudi* malaria in interleukin-4-deficient mice. *Eur J Immunol* **24**: 2285–2293.

Waki, S., Uehara, S., Kanbe, K., Ono, K., Suzuki, M. & Nariuchi, H. (1992) The role of T cells in pathogenesis and protective immunity to murine malaria. *Immunology* **75**: 646–651.

Wassmer, S.C., Cianciolo, G.J., Combes, V. & Grau, G.E. (2005) Inhibition of endothelial activation: a new way to treat cerebral malaria? *Plos Med* **2**: e245.

Waters, A.P., Mota, M.M., van Dijk, M.R. & Janse, C.J. (2005) Parasitology. Malaria vaccines: back to the future? *Science* **307**: 528–530.

Wattavidanage, J., Carter, R., Perera, K.L., Munasingha, A., Bandara, S., McGuinness, D., Wickramasinghe, A.R., Alles, H.K., Mendis, K.N. & Premawansa, S. (1999) TNFalpha*2 marks high risk of severe disease during *Plasmodium falciparum* malaria and other infections in Sri Lankans. *Clin Exp Immunol* **115**: 350–355.

Weatherall, D.J. (2001) Phenotype-genotype relationships in monogenic disease: lessons from the thalassaemias. *Nat Rev Genet* **2**: 245–255.

Weatherall, D.J. & Clegg, J.B. (2001) Inherited haemoglobin disorders: an increasing global health problem. *Bull World Health Organ* **79**: 704–712.

Webster, D. & Hill, A.V. (2003) Progress with new malaria vaccines. *Bull World Health Organ* **81**: 902–909.

Wenisch, C., Parschalk, B., Narzt, E., Looareesuwan, S. & Graninger, W. (1995) Elevated serum levels of IL-10 and IFN-gamma in patients with acute *Plasmodium falciparum* malaria. *Clin Immunol Immunopathol* **74**: 115–117.

Westendorp, R.G., Langermans, J.A., Huizinga, T.W., Elouali, A.H., Verweij, C.L., Boomsma, D.I. & Vandenbroucke, J.P. (1997) Genetic influence on cytokine production and fatal meningococcal disease. *Lancet* **349**: 170–173.

Williams, T.M. (2001) Human leukocyte antigen gene polymorphism and the histocompatibility laboratory. *J Mol Diagn* **3**: 98–104.

Williams, T.N., Mwangi, T.W., Roberts, D.J., Alexander, N.D., Weatherall, D.J., Wambua, S., Kortok, M., Snow, R.W. & Marsh, K. (2005a) An immune basis for malaria protection by the sickle cell trait. *Plos Med* **2**: e128.

Williams, T.N., Wambua, S., Uyoga, S., Macharia, A., Mwacharo, J.K., Newton, C.R. & Maitland, K. (2005b) Both heterozygous and homozygous alpha+ thalassemias protect against severe and fatal *Plasmodium falciparum* malaria on the coast of Kenya. *Blood* **106**: 368–371.

Willimann, K., Matile, H., Weiss, N.A. & Imhof, B.A. (1995) *In vivo* sequestration of *Plasmodium falciparum*-infected human erythrocytes: a severe combined immunodeficiency mouse model for cerebral malaria. *J Exp Med* **182**: 643–653.

Wilson, A.G., Symons, J.A., McDowell, T.L., McDevitt, H.O. & Duff, G.W. (1997) Effects of a polymorphism in the human tumor necrosis factor alpha promoter on transcriptional activation. *Proc Natl Acad Sci USA* **94**: 3195–3199.

Wilson, J.N., Rockett, K., Jallow, M., Pinder, M., Sisay-Joof, F., Newport, M., Newton, J. & Kwiatkowski, D. (2005) Analysis of IL10 haplotypic associations with severe malaria. *Genes Immun* **6**: 462–466.

Wilson, M.E. & Pearson, R.D. (1986) Parasitic diseases of normal hosts in north america. *Hosp Pract (off ed)*, 21, 164a–164d, 164i–164l, 164p–164q pas.

Winograd, E., Prudhomme, J.G. & Sherman, I.W. (2005) Band 3 clustering promotes the exposure of neoantigens in *Plasmodium falciparum*-infected erythrocytes. *Mol Biochem Parasitol* **142**: 98–105.

Winograd, E. & Sherman, I.W. (1989) Characterization of a modified red cell membrane protein expressed on erythrocytes infected with the human malaria parasite *Plasmodium falciparum*: possible role as a cytoadherent mediating protein. *J Cell Biol* **108**: 23–30.

Wong, W.W., Cahill, J.M., Rosen, M.D., Kennedy, C.A., Bonaccio, E.T., Morris, M.J., Wilson, J.G., Klickstein, L.B. & Fearon, D.T. (1989) Structure of the human CR1 gene. Molecular basis of the structural and quantitative polymorphisms and identification of a new CR1-like allele. *J Exp Med* **169**: 847–863.

Yazdani, S.S., Shakri, A.R., Mukherjee, P., Baniwal, S.K. & Chitnis, C.E. (2004) Evaluation of immune responses elicited in mice against a recombinant malaria vaccine based on *Plasmodium vivax* Duffy binding protein. *Vaccine* **22**: 3727–3737.

Yipp, B.G., Anand, S., Schollaardt, T., Patel, K.D., Looareesuwan, S. & Ho, M. (2000) Synergism of multiple adhesion molecules in mediating cytoadherence of *Plasmodium falciparum*-infected erythrocytes to microvascular endothelial cells under flow. *Blood* **96**: 2292–2298.

Yipp, B.G., Baruch, D.I., Brady, C., Murray, A.G., Looareesuwan, S., Kubes, P. & Ho, M. (2003) Recombinant PfEMP1 peptide inhibits and reverses cytoadherence of clinical *Plasmodium falciparum* isolates *in vivo*. *Blood* **101**: 331–337.

Yoneto, T., Waki, S., Takai, T., Tagawa, Y., Iwakura, Y., Mizuguchi, J, Nariuchi, H. & Yoshimoto, T. (2001) A critical role of Fc receptor-mediated antibody-dependent phagocytosis in the host resistance to blood-stage *Plasmodium berghei* XAT infection. *J Immunol* **166**: 6236–6241.

Yuthavong, Y., Butthep, P., Bunyaratvej, A., Fucharoen, S. & Khusmith, S. (1988) Impaired parasite growth and increased susceptibility to phagocytosis of *Plasmodium falciparum* infected alpha-thalassemia or hemoglobin constant spring red blood cells. *Am J Clin Pathol* **89**: 521–525.

Yuthavong, Y., Bunyaratvej, A. & Kamchonwongpaisan, S. (1990) Increased susceptibility of malaria-infected variant erythrocytes to the mononuclear phagocyte system. *Blood Cells* **16**: 591–597.

Zimmerman, P.A., Fitness, J., Moulds, J.M., McNamara, D.T., Kasehagen, L.J., Rowe, J.A. & Hill, A.V. (2003) CR1 Knops blood group alleles are not associated with severe malaria in the Gambia. *Genes Immun* **4**: 368–373.

4

Nicotinic acetylcholine receptors as drug/chemical targets, contributions from comparative genomics, forward and reverse genetics

David B. Sattelle[1], Andrew K. Jones[1], Laurence A. Brown, Steven D. Buckingham, Christopher J. Mee and Luanda Pym

1 Introduction: nicotinic receptor structure and function

Nicotinic acetylcholine receptors (nAChRs) mediate the fast actions of acetylcholine (ACh) in the nervous system and at neuromuscular junctions. They are pentameric, transmembrane proteins containing an integral cation-selective channel (*Figure 1*) (Unwin, 2005). Each of the five subunits has a long N-terminal, extracellular domain containing the ACh-binding site and the dicysteine loop characteristic of ionotropic receptors for ACh, GABA, glycine and serotonin (5-hydoxytryptamine), all of which make up the 'cys-loop' receptor superfamily (Karlin, 2002). The nAChRs can exist as homomers of α subunits, which are defined by a pair of adjacent cysteines in loop C, one of the six loops (A–F) in the extracellular domain that make up the ACh binding site (Corringer *et al.*, 2000). They can also exist as heteromers of either two kinds of α subunit, or, more commonly, of various combinations of α and non-α subunits (Millar, 2003). Although no nAChR crystal structure is currently available, the structure of the *Torpedo marmorata* electric organ nAChR has been resolved at 4.0 Å (Unwin, 2005). In addition, the crystal structure of an acetylcholine binding protein (AChBP) has been determined. The finding that the AChBP shares homology with the extracellular N-terminal region of the nicotinic acetylcholine receptor has facilitated computational modelling of nAChRs (Brejc *et al.*, 2001).

---

[1]These authors have contributed equally to this work.

Comparative Genomics and Proteomics in Drug Discovery, edited by John Parrington and Kevin Coward. © 2007 Taylor and Francis Group.

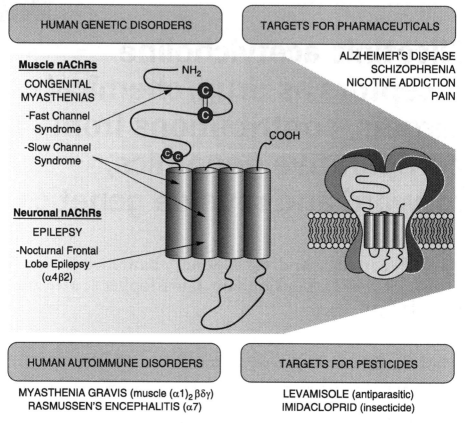

Figure 1. *Nicotinic acetylcholine receptors are composed of five homologous subunits arranged around a central ion channel. Mutations in certain subunits are linked with congenital myasthenia syndromes and one form of epilepsy. Neuronal nAChR subtypes are under investigation as potential targets of therapeutic agents for the control of pain and several disorders. Invertebrate nAChRs are also targets of effective insecticides and anthelmintics.*

2 Nicotinic receptors: roles in human disease

Nicotinic acetylcholine receptors play important roles in genetic and autoimmune disorders and they are targets for important drugs and chemicals (Hogg and Bertrand, 2004).

2.1 Epilepsy

Autosomal dominant nocturnal frontal lobe epilepsy (ADNFLE) is a rare, inherited syndrome that usually occurs during childhood and is characterized by brief, frequent clusters of seizures occurring during light sleep (Combi *et al.*, 2004). This is the only idiopathic epilepsy for which a genetic basis has been established, with three loci being identified so far (chromosomes 20q13, 15q24 and 1q21). The genes corresponding to two of these loci (20q13 and 1q21) are known and they encode the $\alpha4$ and $\beta2$ nAChR subunits. Heterologous expression studies have shown that the mutations of the $\alpha4$ subunit associated with ADNFLE result in impaired receptor function including reduced Ca^{2+} permeability and reduced ACh affinity (Bertrand *et al.*, 1998) as well as shorter channel openings (Figl *et al.*, 1998) and slower desensitization (Matsushima *et al.*, 2002). In contrast, mutations in the β2 subunit appear to exhibit gain of function with increased sensitivity to ACh (Phillips *et al.*, 2001) and faster desensitization (De Fusco *et al.*, 2000).

2.2 *Congenital myasthenias*

Mutations in muscle nAChR subunits underlie congenital myasthenia syndromes (CMS) (Engel *et al.*, 2003), inherited disorders characterized by muscle weakness. Examples of CMS include slow channel syndromes (SCS) and fast channel syndromes (FCS) in which the nicotinic receptor ion channel remains open for longer (SCS) or shorter (FCS) than normal. Proteins that interact with nAChRs also play roles in CMS. For instance, nAChR deficiency disorders can result from reduced expression of nicotinic receptors at the neuromuscular junction as a result of mutations in the nAChR-associated protein rapsyn, which plays a role in receptor clustering (Burke *et al.*, 2003).

2.3 *Myasthenia gravis*

Myasthenia gravis (MG) is an autoimmune disease in which patients generate auto-antibodies to muscle nAChRs resulting in muscle weakness (Hughes *et al.*, 2004). The number of available ACh receptors is reduced not only by the binding of autoanti-bodies specific for nAChRs, but also by deformation of the postsynaptic membrane. Approximately 10% of MG patients lack antibodies to nAChRs but have antibodies to the muscle-specific kinase, MuSK, which plays a role in nAChR clustering and postsynaptic differentiation (Hoch *et al.*, 2001).

2.4 *Rasmussen's encephalitis*

Another autoimmune disease has recently been reported involving nAChRs. Rasmussen's encephalitis is a rare, progressive neurological disorder, the symptoms of which include frequent, severe seizures and loss of motor skills and speech. It is also characterized by paralysis in one side of the body (hemiparesis), encephalitis (brain inflammation), dementia and mental deterioration (Bien *et al.*, 2002). The disorder occurs mainly in children under the age of 15 and was originally thought to be attributed to autoantibodies to the glutamate receptor subunit, GluR3 (Rogers *et al.*, 1994). Recent studies, however, have failed to detect glutamate receptor autoantibodies (Watson *et al.*, 2004). However, the presence of autoantibodies to α7 nAChRs has been detected in some patients and these antibodies block functional α7 receptors (Watson *et al.*, 2005).

3 Nicotinic receptors as drug/chemical targets

3.1 *Human drug targets*

Nicotinic acetylcholine receptors are targets for many new drugs and chemicals. Research is in progress on nicotinic receptors as targets for drugs designed to ameliorate the symptoms of Alzheimer's disease, Parkinson's disease and schizophrenia (Pereira *et al.*, 2002). There is also interest in generating novel analgesics targeting nicotinic receptors (Rashid and Ueda, 2002) and in gaining an improved understanding of nicotine addiction (Graul and Prous, 2005). In the light of a body of evidence that Alzheimer's disease involves a loss of cholinergic transmission (Ballard *et al.*, 2005), many current drug treatments for Alzheimer's disease target the cholinergic system, including anticholinesterases, such as donazepil (Aricept–Eisai Co. Ltd, USA), Rivastigmine (Exelon–Novartis Pharmaceuticals Corporation), galantamine (Razadyne–Ortho-McNeil Neurologics, Inc. – originally extracted from daffodil bulbs) and tacrine (Cognex–Parke-Davis Pharm). Their action is presumed to enhance cholinergic transmission by reducing enzymatic removal of ACh from synapses. The search for possible new targets in the development of Alzheimer's disease treatments may benefit from recent studies that have highlighted the central role of brain $\alpha7$ receptors in both amyloid toxicity and neuroprotection against amyloid toxicity (Chen *et al.*, 2005; Geerts, 2005). Recently, we and others (Grassi *et al.*, 2003) have demonstrated that different receptor subtypes show strikingly different responses to the amyloid peptide ($A\beta_{1-42}$) which forms amyloid plaques in Alzheimer's disease. For example, whereas the brain $\alpha7$ receptor is blocked by this peptide, the response of the $\alpha4\beta2$ receptor to ACh is enhanced. The $\alpha3\beta4$ receptor is unaffected by the application of the $A\beta_{1-42}$ peptide (Pym *et al.*, 2005). Downstream signalling pathways activated by $\alpha7$ receptors may also provide new drug targets, as $A\beta_{1-42}$ has been shown to activate the MAPK pathways in cultured rat hippocampal neurons (Bell *et al.*, 2004). In addition, direct activation of $\alpha7$ nAChRs of cultured neurons by agonists such as nicotine provides protection against amyloid toxicity, through activation of the JNK/MAPK pathway (Shaw *et al.*, 2002).

Nicotinic acetylcholine receptors may also provide new drug targets for the treatment of schizophrenia. Cigarette smoking is more common among schizophrenics than among the general population (McChargue *et al.*, 2002), perhaps being a form of self-medication. Expression of $\alpha7$ receptors, which are associated with auditory sensory gating and smooth eye pursuit – deficits in which are two features of schizophrenia (Deutsch *et al.*, 2005) – is reduced in schizophrenia patients. Similarly, the loss of $\alpha7$ (and other nAChR subtypes) in the brains of Parkinson's disease patients (Burghaus *et al.*, 2003) also points to nAChRs as potential new targets for therapy.

Epibatidine, a potent nAChR antagonist extracted from the poison arrow frog, is 100 times more potent than morphine as an analgesic (Badio and Daly, 1994), but its toxicity and lack of nAChR specificity prohibits its use as a therapeutic agent. However, focal application of nAChR agonists in the brainstem is effective in reducing pain, suggesting that more selective epibatidine analogues might provide new analgesics which, since they do not act through opioid receptors, would be less addictive. ABT-594, an epibatidine analogue, has attracted interest because it retains the analgesic properties with greatly reduced toxicity (Bannon *et al.*, 1998).

3.2 *Control of nematode parasites and insect pests*

A further concern for human health is the devastating impact of parasites and crop pests which reduce harvest yields. Detrimental effects can occur directly by human parasitic infection and by damage to crops and livestock that in turn reduce the availability of food. The nAChRs of parasites and other pests have already proven to be commercially important drug/chemical targets (Harrow and Gration, 1985; Raymond-Delpech *et al.*, 2005).

Helminths (worms and flukes) cause many serious diseases in man and other animals, with the World Health Organization reporting in 1999 that the disease burden was estimated at 40% of all tropical diseases, with about 2000 million people affected worldwide. The cost to the health and economic value of livestock from helminth infection can also be high. Of the available anthelmintic compounds used to control these potentially devastating diseases several have sites of action on nAChRs.

Anthelmintics such as levamisole, pyrantel and morantel have been shown to result in contraction of body wall muscle preparations from the parasitic nematode *Ascaris suum* by activation of nAChRs (Evans and Martin, 1996; Martin *et al.*, 1998). Continued studies of species- and genus-specific properties of parasitic ion channels could give rise to the development of new compounds targeting the nAChRs, as well as a better understanding of the mechanisms whereby resistance to current drugs develops (Kohler, 2001).

With ACh being an abundant neurotransmitter in the nervous systems of many insect species, including the fruitfly, *Drosophila melanogaster* (Baylis *et al.*, 1996; Sattelle and Breer, 1990), nAChRs are targeted by chemicals used for insect control, such as neonicotinoids (Matsuda *et al.*, 2001). Imidacloprid (1-(6-chloro-3-pyridylmethyl)-2-nitroimino-imidazolidine) and other neonicotinoids now have worldwide annual sales of around one billion US dollars, approximately 15% of the global insecticide market (Tomizawa and Casida, 2005).

Radioligand binding and electrophysiological studies have shown that imidacloprid acts on nAChRs in the cockroach central nervous system (Bai *et al.*, 1991), and high affinity binding sites for imidacloprid are present in membrane preparations from diverse insects (Bai *et al.*, 1991; Lind *et al.*, 1998; Liu and Casida, 1993; Zhang *et al.*, 2000). The effectiveness of neonicotinoids as safe insecticides has been attributed, at least in part, to their selectivity for insect nicotinic receptors over mammalian nAChRs (Tomizawa and Casida, 2003) and indeed the binding affinity of neonicotinoids to nAChRs correlates well with insecticidal efficacy (Nishiwaki *et al.*, 2003). Second-generation members of the neonicotinoid family, such as clothianidin, have now been developed (Tomizawa and Casida, 2005).

4 Comparative genomics of nAChR families and their contribution to understanding drug selectivity

4.1 *Mammals*

The mammalian nAChR families have 16 members made up of 10 α and 6 non-α subunits (Millar, 2003). There are separate families of muscle and neuronal nAChRs and the striking differences in pharmacological properties of nicotinic receptor subtypes found in different cells and tissues are mainly attributed to differences in their subunit composition (Hogg *et al.*, 2003).

4.2 *Other vertebrates*

The avian nicotinic receptor family has an additional nAChR subunit, $\alpha 8$ (Schoepfer *et al.*, 1990), which is also present in fish but has not so far been found in mammals. The relatively small family of avian and mammalian nAChR subunits contrasts strikingly with the much larger family (28 subunits) found in the pufferfish, *Fugu rubripes*, which probably arose through genome duplication events (Jones *et al.*, 2003). For instance, there is an increase in the number of muscle subunits in the pufferfish. In addition, there is an expansion of the α9 subfamily of subunits, which has also been observed in the rainbow trout, *Oncorhynchus mykiss* (Drescher *et al.*, 2004). Since α9 is found in cochlear hair cells of mammals (Elgoyhen *et al.*, 1994), it is of interest that in fish there is an even more complex system of hair cells associated with lateral lines all designed for sophisticated mechanoreception. It is also of interest that the β4 α3 α5 triplet cluster of subunits in mammals is in fish a triplet cluster that instead contains β4 α3 and β7 subunits (Jones *et al.*, 2003). This switch between an α and a non-α subunit may have important functional consequences.

4.3 *Invertebrate model organisms: the nematode,* Caenorhabditis elegans, *and the fruitfly,* Drosophila melanogaster

C. elegans possesses the largest nAChR family known with approximately 30 subunits, 27 of which have so far been shown to be transcribed (Jones and Sattelle, 2004). Based on sequence homology, these subunits are divided into five 'core' groups (*Figure 2*) each defined by the first of their number to be discovered (Mongan *et al.*, 1998). The ACR-8 and DEG-3 groups have no counterparts to date in other organisms whilst the ACR-16 group represents a large expansion of subunits highly homologous to the vertebrate α7 nAChR subunit (Mongan *et al.*, 2002). In addition, there are over 20 subunits, denoted orphan subunits, which show substantial homology to nAChRs but do not fall within the five core groups (Jones and Sattelle, 2004). Recently, a subgroup of anion-selective ACh-gated channels has been described, the ACh-binding domains of which have diverged considerably from those of nicotinic receptors (Putrenko *et al.*, 2005). Phylogenetic analysis suggests that they originated from other anion-selective channels such as GABA, glycine and glutamate receptors (*Figure 2*).

   In contrast with *C. elegans*, Drosophila possesses a small nAChR gene family consisting of only 10 subunits (Sattelle *et al.*, 2005). Three of these, Dα5, Dα6 and Dα7, closely resemble vertebrate α7 subunits (Grauso *et al.*, 2002) whilst the other fly subunits do not appear to have obvious analogues in higher organisms.

4.4 *Vectors, pests and beneficial insects*

Sequencing of the *Anopheles gambiae* (malaria mosquito) and *Apis mellifera* (honeybee) genomes has allowed the identification of nAChR gene families from a disease vector/pest and a beneficial insect, respectively (Jones *et al.*, 2005b, 2006). Like Drosophila, the nAChR gene families of these two insects are compact, with the mosquito possessing 10 subunits and the honeybee 11 (*Figure 3*). The mosquito and honeybee orthologues of Dβ2 are α subunits (Agamα8 and Amelα8, respectively) and, interestingly, these subunits form part of nAChR subunit gene clusters mirroring the α and non-α switch observed in the pufferfish (Jones *et al.*, 2003). It is worth noting that Drosophila, Anopheles and Apis possess at least one nAChR subunit that is highly divergent not only to nematode and vertebrate nAChR subunits but within insect species. Thus, Drosophila has one divergent nAChR subunit which is non-α (Dβ3 [Lansdell and Millar, 2002]), Anopheles also possesses one divergent subunit, but it is an α (Agamα9 [Jones *et al.*, 2005b]) whilst there are two divergent subunits in

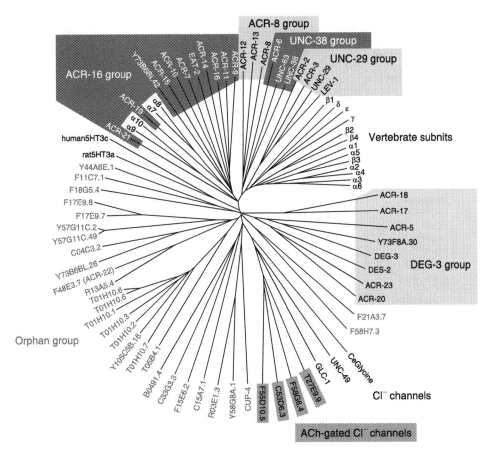

Figure 2. *The ACh-gated ion channel superfamily of* C. elegans. *The nAChR subunits are divided into five core groups whilst subunits showing substantial sequence homology with known nAChR subunits that do not fall within the core groups are designated 'Orphan' subunits. ACh-gated chloride channels show greatest similarity to other anion channels such as GABA, glycine and glutamate receptors. The avian/mammalian nAChR gene family is also included for comparison.*

apis, one an α subunit, the other non-α (Amel9α and Amelβ2 [Jones *et al.*, 2006]). The genome sequencing of the red flour beetle, *Tribolium castaneum*, is close to completion (http://www.hgsc.bcm.tmc.edu/projects/tribolium/), which will allow the characterization of an insect pest nAChR gene family.

5 Splicing and RNA editing in nicotinic acetylcholine receptors adds to diversity

Even though insects possess a small number of nAChR subunit genes, alternative splicing (Stetefeld and Ruegg, 2005) and pre-mRNA A-to-I editing (Seeburg, 2002)

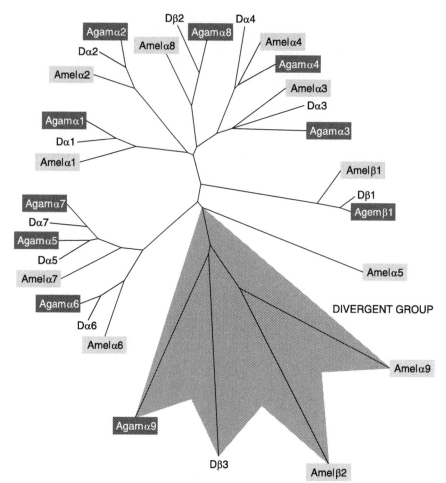

Figure 3. *The nAChR ion channel gene families of* Drosophila melanogaster, Anopheles gambiae *and* Apis mellifera*. In addition to possessing highly conserved subunits, each insect species has at least one highly divergent subunit.*

vastly increases the nicotinic receptor proteome (Grauso *et al.*, 2002; Hoopengardner *et al.*, 2003; Sattelle *et al.*, 2005). Four Drosophila nAChR subunits are alternatively spliced resulting in either exon substitution or subunit truncation (*Figure 4A*). The pattern of splicing in the equivalent subunits in Anopheles and Apis is similar to that of Drosophila (Jones *et al.*, 2005b, 2006). In addition, four Drosophila subunits undergo RNA editing (*Figure 4B*) although no RNA editing was detected in Anopheles nAChR subunits (Jones *et al.*, 2005b) and only one Apis subunit (Amelα6) is edited (Jones *et al.*, 2006). However, like its Drosophila orthologue Dα6, Amelα6 is edited at several sites clustered within the N-terminal extracellular domain (*Figure 4B*), only two of which are conserved between both species. Thus, RNA editing serves to not only broaden the insect nAChR proteome, but also to produce species-specific sub-unit variants.

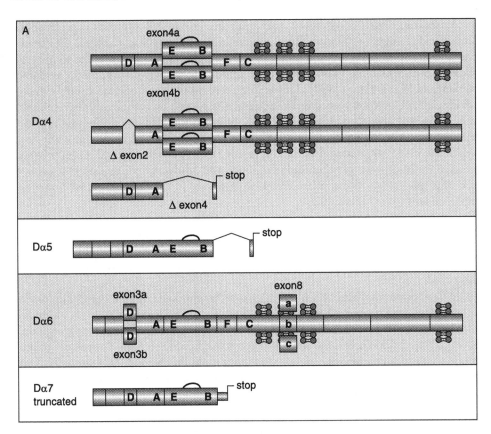

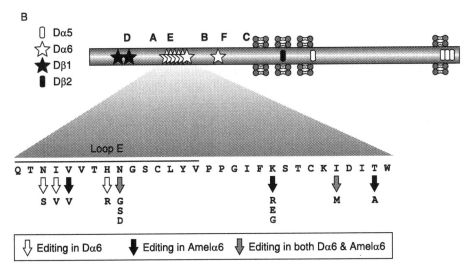

Figure 4. (A) Alternative splicing of Drosophila nAChR subunits. The ligand-binding domains (loops A-F) are indicated as well as the cys-loop (black bar located between loops E and B) and the four transmembrane regions. (B) RNA editing of insect nAChR subunits. Known RNA editing sites of Drosophila nAChR subunits are shown on a schematic of a nAChR subunit and a comparison of editing between Dα6 and its honeybee orthologue, Amelα6, is included.

Alternative splicing and RNA editing of insect nAChR subunits can introduce amino acid changes in functionally significant regions (Sattelle *et al.*, 2005). For example, alternative splicing of exon 4 of Dα4 alters residues within, or in the vicinity of, the cys-loop, which has been shown to be important for complete receptor assembly (Green and Wanamaker, 1997). This is in agreement with radioligand-binding assays which indicate that subunits containing exon 4′ assemble less efficiently than subunits with exon 4 (Lansdell and Millar, 2000a). Interestingly, RT-PCR analysis revealed that the two Apis α4 splice variants are differentially expressed throughout the honeybee life cycle with exon 4 variants present at each developmental stage, whereas exon 4′ variants were detected only in pupae and adults. This suggests that exon 4′ subunits may serve to modulate receptor assembly in the later stages of honeybee development (Jones *et al.*, 2006).

The function, if any, of the truncated nAChR subunits has yet to be determined but it has been suggested that, if they are capable of binding ACh, they may act as an 'ACh sponge' serving to terminate cholinergic transmission in a manner similar to that of the molluscan ACh-binding protein (Grauso *et al.*, 2002; Sattelle *et al.*, 2005; Smit *et al.*, 2003).

In vertebrate nAChRs, RNA editing has so far not been observed and differential splicing has been shown where a variant of the rodent α7 subunit possesses a stretch of 27 amino acids incorporated into the N-terminal ligand binding region (Severance *et al.*, 2004). Like α7, this variant (denoted α7-2) is able to form functional homomeric channels although with higher sensitivity to ACh, altered α-bungarotoxin binding properties and a slower rate of desensitization. A similar variant for the human α7 subunit has so far not been reported.

6 Forward genetics in the study of *C. elegans* nicotinic receptor subunit function

With the difficulties of culturing parasitic nematodes outside a host, the genetic model organism, *C. elegans*, which can be easily grown and studied in the laboratory, has been utilized in elucidating the targets of anthelmintics. In a classic example using *C. elegans*, a forward genetics approach, in the form of a chemistry-to-gene genetic screen, has identified nAChR subunits that make up the target of the antiparasitic drug levamisole (Brenner, 1974; Jones *et al.*, 2005a; Lewis *et al.*, 1980, 1987). These subunits (UNC-63, UNC-38, LEV-8, UNC-29 and LEV-1), all of which are expressed in the body wall muscle, were detected in this genetic screen by either their resistance to levamisole or uncoordinated movement (Brenner, 1974; Culetto *et al.*, 2004; Fleming *et al.*, 1997; Towers *et al.*, 2005). The genetic screen for factors underlying levamisole resistance also identified important gene products acting upstream and downstream of nAChRs (Jones *et al.*, 2005a). These include: LEV-10 which is required for nAChR aggregation at the neuromuscular junction (Gally *et al.*, 2004); UNC-50 which is involved in the processing and assembly of receptors (Lewis *et al.*, 1987), the mammalian homologue of which was subsequently identified and found to perform a similar role (Fitzgerald *et al.*, 2000); LEV-11 and UNC-22 which regulate muscle contraction (Benian *et al.*, 1993; Kagawa *et al.*, 1997); and UNC-68 (a ryanodine receptor) which is involved in calcium signalling (Maryon *et al.*, 1996). Since the ryanodine receptor is now known to be the target of anthranilic diamides, an important new class of chemicals targeting invertebrate pests (Lahm *et al.*, 2005), this chemistry-to-gene screen approach can be seen to isolate novel drug/chemical targets. Another chemistry-to-gene screen investigating factor underlying resistance to the pesticide aldicarb, an acetylcholinesterase inhibitor, identified other proteins associated with nAChR function (for review, see Jones *et al.*, 2005a), most notably RIC-3, which is involved in receptor assembly and trafficking (Halevi *et al.*, 2002).

A gain of function mutation in the DEG-3 nAChR subunit results in increased calcium permeability, causing neuronal degeneration (Treinin and Chalfie, 1995). The *deg-3* gene is part of an operon (Blumenthal *et al.*, 2002), together with another nAChR subunit, *des-2*, indicating coordinate expression (Treinin *et al.*, 1998; Yassin *et al.*, 2001). Indeed, both subunits co-assemble in Xenopus oocytes to form functional channels highly sensitive to choline and they show similar expression patterns, which include chemosensory neurons, suggesting a role in chemotaxis. In line with this, homozygous mutations in *deg-3* mutants are defective in chemotaxis towards choline.

Members of the large orphan group of *C. elegans* nAChR candidate subunits (*Figure 2*) remained uncharacterized until recently when CUP-4 and Y58G8A.1 were found to be required for efficient endocytosis of fluids by coelomocytes (Patton *et al.*, 2005). This, together with the widely expressed *lev-8* (Towers *et al.*, 2005), highlights the fact that many nAChRs are expressed in non-neuronal cells. There are also examples from vertebrates, such as α7, which is expressed in muscle (Corriveau *et al.*, 1995), endothelial cells (Moccia *et al.*, 2004) and skin (Valiante *et al.*, 2004). Thus, nAChRs subunits can engage in a wide variety of processes other than synaptic transmission. In one example, the cholinergic system can exert an anti-inflammatory effect via nAChR expressed in macrophages offering novel exciting therapeutic potential of nicotinic receptors as targets for anti-inflammatory drugs (Ulloa, 2005).

7 Reverse genetics in the functional analysis of nicotinic acetylcholine receptors

With the advent of RNA interference (RNAi) (Fire *et al.*, 1998), reverse genetics, an approach that starts with silencing genes of known sequence and observing resulting phenotypes, has been widely used to study gene function. The fully sequenced genome and the amenability of the organism to RNAi studies has allowed *C. elegans* to be used in genome-wide RNAi screens which have uncovered new gene functions (Kamath *et al.*, 2003; Simmer *et al.*, 2003). The use of RNAi to silence more than one gene simultaneously can be exploited to overcome functional redundancy between genes and similar results can be achieved by applying RNAi to silence a single gene in a mutant background. The latter approach was employed successfully to identify a second subtype of nAChR in the *C. elegans* body wall muscle that is insensitive to levamisole (Culetto *et al.*, 2004; Richmond and Jorgensen, 1999). It was found that RNAi targeting *acr-16* substantially reduced motility of worms having a null mutation in *unc-29* (a levamisole-sensitive nAChR subunit) whilst silencing of *acr-16* in wild-type worms had no effect (Francis *et al.*, 2005). Subsequent electrophysiological studies on body wall muscles showed that ACR-16 is not a component of the levamisole-sensitive receptor but is essential for nicotine responses. Since levamisole demonstrated that body wall muscle nAChRs are effective drug targets, ACR-16 may represent an important novel target for parasitic nematode control. Another study, using microarray analysis of transcripts enriched in *C. elegans* muscle, also identified ACR-16 as a component of the body wall muscle levamisole-insensitive receptor (Touroutine *et al.*, 2005). In addition, a further body wall muscle nAChR subunit (ACR-8) was identified, although it does not appear to be an essential component of either the levamisole-sensitive or levamisole-insensitive receptor.

RNAi can also be successfully applied to *Drosophila melanogaster*. In GFP-labelled, cultured cholinergic neurons from Drosophila larvae, the importance of the Dα2 (SAD) nAChR subunit in functional responses to nicotine has been established using RNAi combined with ratiometric calcium imaging (Brown *et al.*, 2003). The presence of the Dα2 subunit is supported by the results of microarray experiments carried out on dissociated larval neurons from the same line of flies, following fluorescent activated cell sorting (FACS). Expanding this approach will provide further insights into the roles and functional characteristics of individual subunit members of the nAChR family.

8 Heterologous expression studies on nAChRs provide insights into selectivity of neonicotinoids for insects over vertebrates

Whilst there is an example of a functional insect nAChR expressed in Xenopus oocytes (Jones *et al.*, 2005c; Marshall *et al.*, 1990), the heterologous expression of functionally robust nAChRs still remains elusive. Nevertheless, Drosophila nAChR α subunits can form robust functional receptors when co-expressed with a vertebrate β2 subunit (Bertrand *et al.*, 1994; Sattelle *et al.*, 2005). Expression studies have shown that the Dα1/β2 and Dα2/β2 hybrid nAChRs are more sensitive to imidacloprid than the vertebrate α4β2 receptor (Ihara *et al.*, 2003; Matsuda *et al.*, 1998), suggesting that both Drosophila α subunits have structural features favourable for selective interactions with neonicotinoids. Deleting an insertion between loop B and C in Dα2, which is not found in vertebrate subunits, and replacing a proline in loop C, which is particular to insect nAChRs, with an acidic residue more common in vertebrate subunits significantly reduced Dα2/β2 hybrid sensitivity to imidacloprid, highlighting insect-specific features giving rise to neonicotinoid selectivity (Shimomura *et al.*, 2004). Also, the use of chimeras replacing regions of α4 with portions of Dα2 indicate that loop B to the N-terminus in Dα2 contributes to high imidacloprid sensitivity (Shimomura *et al.*, 2005). The difficulty of expressing functional receptors with Drosophila β subunits have prevented similar investigations into the interactions of non-α subunits with insecticides. However, recombinant hybrid nAChRs stably expressed in cell lines have shown that binding of imidacloprid to the Dα3/β2 receptor was abolished by replacing the β2 with β4, suggesting that non-α subunits may play a role in determining neonicotinoid sensitivity (Lansdell and Millar, 2000b).

Site-directed mutagenesis applied to loop F has shown that G189D and G189E mutations in chicken α7 expressed in Xenopus oocytes markedly reduce receptor responses to neonicotinoids whereas G189N and G189Q mutations scarcely influence them (Matsuda *et al.*, 2000), reinforcing the importance of electrostatic interactions for imidacloprid binding (Matsuda *et al.*, 2005). A similar site-directed mutagenesis approach in loop D indicates that basic residues present in insect but not vertebrate nAChR subunits are likely to contribute to the selective actions of neonicotinoids (Shimomura *et al.*, 2002).

9 Conclusions: lessons from comparative genomics of nicotinic acetylcholine receptor families

The diversity of nAChR subtypes is achieved by multiple subunits combining in various ways. For example, different combinations of vertebrate neuronal subunits result in 30 described subtypes (Millar, 2003) and organisms with larger nAChR gene families (e.g. nematode and pufferfish) have the potential for an even greater range of functional subtypes. Extra subunit diversity is brought about by alternative splicing and RNA editing (particularly in insects) so that apparently small insect nAChR gene families are in reality much more complex (Sattelle et al., 2005). Such species-specific proteome diversification, as well as the presence of highly divergent subunits, represents promising differences to target for future rational insecticide design and improving insecticide selectivity for pest over beneficial insects (e.g. honeybee). Also, identifying the diversity of nAChR subtypes particular to insect and nematode species should provide even further opportunities for the targeting of parasitic species and reducing the side effects of antiparasitic drugs on the host. Studies of heterologously expressed hybrid insect nAChRs have shed light on imidacloprid selectivity for insect over mammalian receptors. With the emergence of levamisole and imidacloprid resistance (Liu et al., 2005; Robertson et al., 1999), future expression studies, in combination with molecular modelling, may prove instructive in discerning the basis of pesticide resistance.

The use of C. elegans has identified from a large nAChR gene family a subset of subunits targeted by the antiparasitic drug, levamisole, as well as shedding light on nAChR-associated proteins found not just in nematodes, but also humans (Jones et al., 2005a). A recent study using tandem affinity purification of the C. elegans levamisole-sensitive receptor (Gottschalk et al., 2005) identified additional nAChR-associated proteins such as: the calcineurin A subunit, TAX-6; copine NRA-1; and SOC-1 which acts in receptor tyrosine kinase signalling. Proteins associated with nAChR function are of interest in the light of the finding that rapsyn mutations underlie several forms of CMS (Burke et al., 2003). In future, other nAChR-associated proteins may also be found to play a role in genetic disorders.

References

Badio, B. and Daly, J.W. (1994) Epibatidine, a potent analgetic and nicotinic agonist. Mol Pharmacol 45: 563–569.

Bai, D.L., Lummis, S.C.R., Leicht, W., Breer, H. and Sattelle, D.B. (1991) Actions of imidacloprid and a related nitromethylene on cholinergic receptors of an identified insect motor-neuron. Pesticide Sci 33: 197–204.

Ballard, C.G., Greig, N.H., Guillozet-Bongaarts, A.L., Enz, A. and Darvesh, S. (2005) Cholinesterases: roles in the brain during health and disease. Curr Alzheimer Res 2: 307–318.

Bannon, A.W., Decker, M.W., Curzon, P., Buckley, M.J., Kim, D.J., Radek, R.J., Lynch, J.K., Wasicak, J.T., Lin, N.H., Arnold, W.H., Holladay, M.W., Williams, M. and Arneric, S.P. (1998) ABT-594 [(R)-5-(2-azetidinylmethoxy)-2-chloropyridine]: a novel, orally effective antinociceptive agent acting via neuronal nicotinic acetylcholine receptors: II. In vivo characterization. J Pharmacol Exp Ther 285: 787–794.

Baylis, H.A., Sattelle, D.B. and Lane, N.J. (1996) Genetic analysis of cholinergic nerve terminal function in invertebrates. *J Neurocytol* **25**: 747–762.

Bell, K.A., O'Riordan, K.J., Sweatt, J.D. and Dineley, K.T. (2004) MAPK recruitment by beta-amyloid in organotypic hippocampal slice cultures depends on physical state and exposure time. *J Neurochem* **91**: 349–361.

Benian, G.M., L'Hernault, S.W. and Morris, M.E. (1993) Additional sequence complexity in the muscle gene, unc-22, and its encoded protein, twitchin, of *Caenorhabditis elegans*. *Genetics* **134**: 1097–1104.

Bertrand, D., Ballivet, M., Gomez, M., Bertrand, S., Phannavong, B. and Gundelfinger, E.D. (1994) Physiological properties of neuronal nicotinic receptors reconstituted from the vertebrate beta 2 subunit and Drosophila alpha subunits. *Eur J Neurosci* **6**: 869–875.

Bertrand, S., Weiland, S., Berkovic, S.F., Steinlein, O.K. and Bertrand, D. (1998) Properties of neuronal nicotinic acetylcholine receptor mutants from humans suffering from autosomal dominant nocturnal frontal lobe epilepsy. *Br J Pharmacol* **125**: 751–760.

Bien, C.G., Widman, G., Urbach, H., Sassen, R., Kuczaty, S., Wiestler, O.D., Schramm, J. and Elger, C.E. (2002) The natural history of Rasmussen's encephalitis. *Brain* **125**: 1751–1759.

Blumenthal, T., Evans, D., Link, C.D., Guffanti, A., Lawson, D., Thierry-Mieg, J., Thierry-Mieg, D., Chiu, W.L., Duke, K., Kiraly, M. and Kim, S.K. (2002) A global analysis of *Caenorhabditis elegans* operons. *Nature* **417**: 851–854.

Brejc, K., van Dijk, W.J., Klaassen, R.V., Schuurmans, M., van Der Oost, J., Smit, A.B. and Sixma, T.K. (2001) Crystal structure of an ACh-binding protein reveals the ligand-binding domain of nicotinic receptors. *Nature* **411**: 269–276.

Brenner, S. (1974) The genetics of *Caenorhabditis elegans*. *Genetics* **77**: 71–94.

Brown, L.A., Jepson, J.J., Song, S., Salvaterra, P.M. and Sattelle, D.B. (2003) Functional genomics of the *Drosophila melanogaster* nicotinic acetylcholine receptor gene family. In: Beadle, D., Mellor, I. and Usherwood, P.N.R. (eds) *Neurotox 2003: Neurotoxicological Targets from Functional Genomics and Proteomics*. Society for Chemical Industry, London: 131–139.

Burghaus, L., Schutz, U., Krempel, U., Lindstrom, J. and Schroder, H. (2003) Loss of nicotinic acetylcholine receptor subunits alpha4 and alpha7 in the cerebral cortex of Parkinson patients. *Parkinsonism Relat Disord* **9**: 243–246.

Burke, G., Cossins, J., Maxwell, S., Owens, G., Vincent, A., Robb, S., Nicolle, M., Hilton-Jones, D., Newsom-Davis, J., Palace, J. and Beeson, D. (2003) Rapsyn mutations in hereditary myasthenia: distinct early- and late-onset phenotypes. *Neurology* **61**: 826–828.

Chen, L., Yamada, K., Nabeshima, T. and Sokabe, M. (2005) alpha7 Nicotinic acetylcholine receptor as a target to rescue deficit in hippocampal LTP induction in beta-amyloid infused rats. *Neuropharmacology* **50**: 254–268.

Combi, R., Dalpra, L., Tenchini, M.L. and Ferini-Strambi, L. (2004) Autosomal dominant nocturnal frontal lobe epilepsy – a critical overview. *J Neurol* **251**: 923–934.

Corringer, P.J., Le Novere, N. and Changeux, J.P. (2000) Nicotinic receptors at the amino acid level. *Annu Rev Pharmacol Toxicol* **40**: 431–458.

Corriveau, R.A., Romano, S.J., Conroy, W.G., Oliva, L. and Berg, D.K. (1995) Expression of neuronal acetylcholine receptor genes in vertebrate skeletal muscle during development. *J Neurosci* 15: 1372–1383.

Culetto, E., Baylis, H.A., Richmond, J.E., Jones, A.K., Fleming, J.T., Squire, M.D., Lewis, J.A. and Sattelle, D.B. (2004) The *Caenorhabditis elegans* unc-63 gene encodes a levamisole-sensitive nicotinic acetylcholine receptor alpha subunit. *J Biol Chem* 279: 42476–42483.

De Fusco, M., Becchetti, A., Patrignani, A., Annesi, G., Gambardella, A., Quattrone, A., Ballabio, A., Wanke, E. and Casari, G. (2000) The nicotinic receptor beta 2 subunit is mutant in nocturnal frontal lobe epilepsy. *Nat Genet* 26: 275–276.

Deutsch, S.I., Rosse, R.B., Schwartz, B.L., Weizman, A., Chilton, M., Arnold, D.S. and Mastropaolo, J. (2005) Therapeutic implications of a selective alpha7 nicotinic receptor abnormality in schizophrenia. *Isr J Psychiatry Relat Sci* 42: 33–44.

Drescher, D.G., Ramakrishnan, N.A., Drescher, M.J., Chun, W., Wang, X., Myers, S.F., Green, G.E., Sadrazodi, K., Karadaghy, A.A., Poopat, N., Karpenko, A.N., Khan, K.M. and Hatfield, J.S. (2004) Cloning and characterization of alpha9 subunits of the nicotinic acetylcholine receptor expressed by saccular hair cells of the rainbow trout (*Oncorhynchus mykiss*). *Neuroscience* 127: 737–752.

Elgoyhen, A.B., Johnson, D.S., Boulter, J., Vetter, D.E. and Heinemann, S. (1994) Alpha 9: an acetylcholine receptor with novel pharmacological properties expressed in rat cochlear hair cells. *Cell* 79: 705–715.

Engel, A.G., Ohno, K. and Sine, S.M. (2003) Congenital myasthenic syndromes: progress over the past decade. *Muscle Nerve* 27: 4–25.

Evans, A.M. and Martin, R.J. (1996) Activation and cooperative multi-ion block of single nicotinic-acetylcholine channel currents of Ascaris muscle by the tetrahydropyrimidine anthelmintic, morantel. *Br J Pharmacol* 118: 1127–1140.

Figl, A., Viseshakul, N., Shafaee, N., Forsayeth, J. and Cohen, B.N. (1998) Two mutations linked to nocturnal frontal lobe epilepsy cause use-dependent potentiation of the nicotinic ACh response. *J Physiol* 513: 655–670.

Fire, A., Xu, S., Montgomery, M.K., Kostas, S.A., Driver, S.E. and Mello, C.C. (1998) Potent and specific genetic interference by double-stranded RNA in *Caenorhabditis elegans*. *Nature* 391: 806–811.

Fitzgerald, J., Kennedy, D., Viseshakul, N., Cohen, B.N., Mattick, J., Bateman, J.F. and Forsayeth, J.R. (2000) UNCL, the mammalian homologue of UNC-50, is an inner nuclear membrane RNA-binding protein. *Brain Res* 877: 110–123.

Fleming, J.T., Squire, M.D., Barnes, T.M., Tornoe, C., Matsuda, K., Ahnn, J., Fire, A., Sulston, J.E., Barnard, E.A., Sattelle, D.B. and James, L.A. (1997) *Caenorhabditis elegans* levamisole resistance genes *lev-1*, *unc-29*, and *unc-38* encode functional acetylcholine receptor subunits. *J Neurosci* 15: 5843–5857.

Francis, M.M., Evans, S.P., Jensen, M., Madsen, D.M., Mancuso, J., Norman, K.R. and Maricq, A.V. (2005) The Ror receptor tyrosine kinase CAM-1 is required for ACR-16-mediated synaptic transmission at the *C. elegans* neuromuscular junction. *Neuron* 46: 581–594.

Gally, C., Eimer, S., Richmond, J.E. and Bessereau, J.L. (2004) A transmembrane protein required for acetylcholine receptor clustering in *Caenorhabditis elegans*. *Nature* 431: 578–582.

Geerts, H. (2005) Indicators of neuroprotection with galantamine. *Brain Res Bull* **64**: 519–524.

Gottschalk, A., Almedom, R.B., Schedletzky, T., Anderson, S.D., Yates, J.R., 3rd and Schafer, W.R. (2005) Identification and characterization of novel nicotinic receptor-associated proteins in *Caenorhabditis elegans*. *Embo J* **24**: 2566–2578.

Grassi, F., Palma, E., Tonini, R., Amici, M., Ballivet, M. and Eusebi, F. (2003) Amyloid beta(1–42) peptide alters the gating of human and mouse alpha-bungarotoxin-sensitive nicotinic receptors. *J Physiol* **547**: 147–157.

Graul, A.I. and Prous, J.R. (2005) Executive summary: nicotine addiction. *Drugs Today (Barc)* **41**: 419–425.

Grauso, M., Reenan, R.A., Culetto, E. and Sattelle, D.B. (2002) Novel putative nicotinic acetylcholine receptor subunit genes, Dalpha5, Dalpha6 and Dalpha7, in *Drosophila melanogaster* identify a new and highly conserved target of adenosine deaminase acting on RNA-mediated A-to-I pre-mRNA editing. *Genetics* **160**: 1519–1533.

Green, W.N. and Wanamaker, C.P. (1997) The role of the cystine loop in acetylcholine receptor assembly. *J Biol Chem* **272**: 20945–20953.

Halevi, S., McKay, J., Palfreyman, M., Yassin, L., Eshel, M., Jorgensen, E. and Treinin, M. (2002) The *C. elegans* ric-3 gene is required for maturation of nicotinic acetylcholine receptors. *Embo J* **21**: 1012–1020.

Harrow, I.D. and Gration, K.A.F. (1985) Mode of action of the anthelmintics morantel, pyrantel and levamisole on muscle cell membrane of the nematode *Ascaris Suum*. *Pestic Sci* **16**: 662–672.

Hoch, W., McConville, J., Helms, S., Newsom-Davis, J., Melms, A. and Vincent, A. (2001) Auto-antibodies to the receptor tyrosine kinase MuSK in patients with myasthenia gravis without acetylcholine receptor antibodies. *Nat Med* **7**: 365–368.

Hogg, R.C. and Bertrand, D. (2004) Nicotinic acetylcholine receptors as drug targets. *Curr Drug Targets CNS Neurol Disord* **3**: 123–130.

Hogg, R.C., Raggenbass, M. and Bertrand, D. (2003) Nicotinic acetylcholine receptors: from structure to brain function. *Rev Physiol Biochem Pharmacol* **147**: 1–46.

Hoopengardner, B., Bhalla, T., Staber, C. and Reenan, R. (2003) Nervous system targets of RNA editing identified by comparative genomics. *Science* **301**: 832–836.

Hughes, B.W., Moro De Casillas, M.L. and Kaminski, H.J. (2004) Pathophysiology of myasthenia gravis. *Semin Neurol* **24**: 21–30.

Ihara, M., Matsuda, K., Otake, M., Kuwamura, M., Shimomura, M., Komai, K., Akamatsu, M., Raymond, V. and Sattelle, D.B. (2003) Diverse actions of neonicotinoids on chicken alpha7, alpha4beta2 and Drosophila-chicken SADbeta2 and ALSbeta2 hybrid nicotinic acetylcholine receptors expressed in *Xenopus laevis* oocytes. *Neuropharmacology* **45**: 133–144.

Jones, A.K. and Sattelle, D.B. (2003) Functional genomics of the nicotinic acetylcholine receptor gene family of the nematode, *Caenorhabditis elegans*. *Bioessays* **26**: 39–49.

Jones, A.K., Elgar, G. and Sattelle, D.B. (2003) The nicotinic acetylcholine receptor gene family of the pufferfish, *Fugu rubripes*. *Genomics* **82**: 441–451.

Jones, A.K., Buckingham, S.D. and Sattelle, D.B. (2005a) Chemistry-to-gene screens in *Caenorhabditis elegans*. *Nat Rev Drug Discov* **4**: 321–330.

Jones, A.K., Grauso, M. and Sattelle, D.B. (2005b) The nicotinic acetylcholine receptor gene family of the malaria mosquito, *Anopheles gambiae. Genomics* **85**: 176–187.

Jones, A.K., Marshall, J., Blake, A.D., Buckingham, S.D., Darlison, M.G. and Sattelle, D.B. (2005c) Sgbeta1, a novel locust (*Schistocerca gregaria*) non-alpha nicotinic acetylcholine receptor-like subunit with homology to the *Drosophila melanogaster* Dbeta1 subunit. *Invert Neurosci* **5**: 147–155.

Jones, A.K., Raymond-Delpech, V., Thany, S.H., Gauthier, M. and Sattelle, D.B. (2006) The nicotinic acetylcholine receptor gene family of the honeybee, *Apis mellifera. Gen Res* **In press.**

Kagawa, H., Takuwa, K. and Sakube, Y. (1997) Mutations and expressions of the tropomyosin gene and the troponin C gene of *Caenorhabditis elegans. Cell Struct Funct* **22**: 213–218.

Kamath, R.S., Fraser, A.G., Dong, Y., Poulin, G., Durbin, R., Gotta, M., Kanapin, A., Le Bot, N., Moreno, S., Sohrmann, M., Welchman, D.P., Zipperlen, P. and Ahringer, J. (2003) Systematic functional analysis of the *Caenorhabditis elegans* genome using RNAi. *Nature* **421**: 231–237.

Karlin, A. (2002) Emerging structure of the nicotinic acetylcholine receptors. *Nat Rev Neurosci* **3**: 102–114.

Kohler, P. (2001) The biochemical basis of anthelmintic action and resistance. *Int J Parasitol* **31**: 336–345.

Lahm, G.P., Selby, T.P., Freudenberger, J.H., Stevenson, T.M., Myers, B.J., Seburyamo, G., Smith, B.K., Flexner, L., Clark, C.E. and Cordova, D. (2005) Insecticidal anthranilic diamides: a new class of potent ryanodine receptor activators. *Bioorg Med Chem Lett* **15**: 4898–4906.

Lansdell, S.J. and Millar, N.S. (2000a) Cloning and heterologous expression of Dalpha4, a *Drosophila* neuronal nicotinic acetylcholine receptor subunit: identification of an alternative exon influencing the efficiency of subunit assembly. *Neuropharmacology* **39**: 2604–2614.

Lansdell, S.J. and Millar, N.S. (2000b) The influence of nicotinic receptor subunit composition upon agonist, alpha-bungarotoxin and insecticide (imidacloprid) binding affinity. *Neuropharmacology* **39**: 671–679.

Lansdell, S.J. and Millar, N.S. (2002) Dbeta3, an atypical nicotinic acetylcholine receptor subunit from Drosophila: molecular cloning, heterologous expression and coassembly. *J. Neurochem.* **80**: 1009–1018.

Lewis, J.A., Wu, C.H., Levine, J.H. and Berg, H. (1980) Levamisole-resistant mutants of the nematode *Caenorhabditis elegans* appear to lack pharmacological acetylcholine receptors. *Neuroscience* **5**: 967–989.

Lewis, J.A., Elmer, J.S., Skimming, J., McLafferty, S., Fleming, J. and McGee, T. (1987) Cholinergic receptor mutants of the nematode *Caenorhabditis elegans. J Neurosci* **7**: 3059–3071.

Lind, R.J., Clough, M.S., Reynolds, S.E. and Earley, F.G.P. (1998) H-3 imidacloprid labels high- and low-affinity nicotinic acetylcholine receptor-like binding sites in the aphid *Myzus persicae* (Hemiptera: Aphididae). *Pesticide Biochem Physiol* **62**: 3–14.

Liu, M.Y. and Casida, J.E. (1993) High-affinity binding of H-3 imidacloprid in the insect acetylcholine-receptor. *Pesticide Biochem Physiol* **46**: 40–46.

Liu, Z., Williamson, M.S., Lansdell, S.J., Denholm, I., Han, Z. and Millar, N.S. (2005) A nicotinic acetylcholine receptor mutation conferring target-site resistance to imidacloprid in *Nilaparvata lugens* (brown planthopper). *Proc Natl Acad Sci USA* 102: 8420–8425.

Marshall, J., Buckingham, S.D., Shingai, R., Lunt, G.G., Goosey, M.W., Darlison, M.G., Sattelle, D.B. and Barnard, E.A. (1990) Sequence and functional expression of a single alpha subunit of an insect nicotinic acetylcholine receptor. *Embo J* 9: 4391–4398.

Martin, R.J., Murray, I., Robertson, A.P., Bjorn, H. and Sangster, N. (1998) Anthelmintics and ion-channels: after a puncture, use a patch. *Int J Parasitol* 28: 849–862.

Maryon, E.B., Coronado, R. and Anderson, P. (1996) unc-68 encodes a ryanodine receptor involved in regulating *C. elegans* body-wall muscle contraction. *J Cell Biol* 134: 885–893.

Matsuda, K., Buckingham, S.D., Freeman, J.C., Squire, M.D., Baylis, H.A. and Sattelle, D.B. (1998) Effects of the alpha subunit on imidacloprid sensitivity of recombinant nicotinic acetylcholine receptors. *Br. J. Pharmacol* 123: 518–524.

Matsuda, K., Shimomura, M., Kondo, Y., Ihara, M., Hashigami, K., Yoshida, N., Raymond, V., Mongan, N.P., Freeman, J.C., Komai, K. and Sattelle, D.B. (2000) Role of loop D of the alpha7 nicotinic acetylcholine receptor in its interaction with the insecticide imidacloprid and related neonicotinoids. *Br J Pharmacol* 130: 981–986.

Matsuda, K., Buckingham, S.D., Kleier, D., Rauh, J.J., Grauso, M. and Sattelle, D.B. (2001) Neonicotinoids: insecticides acting on insect nicotinic acetylcholine receptors. *Trends Pharmacol Sci* 22: 573–580.

Matsuda, K., Shimomura, M., Ihara, M., Akamatsu, M. and Sattelle, D.B. (2005) Neonicotinoids show selective and diverse actions on their nicotinic receptor targets: electrophysiology, molecular biology, and receptor modeling studies. *Biosci Biotechnol Biochem* 69: 1442–1452.

Matsushima, N., Hirose, S., Iwata, H., Fukuma, G., Yonetani, M., Nagayama, C., Hamanaka, W., Matsunaka, Y., Ito, M., Kaneko, S., Mitsudome, A. and Sugiyama, H. (2002) Mutation (Ser284Leu) of neuronal nicotinic acetylcholine receptor alpha 4 subunit associated with frontal lobe epilepsy causes faster desensitization of the rat receptor expressed in oocytes. *Epilepsy Res* 48: 181–186.

McChargue, D.E., Gulliver, S.B. and Hitsman, B. (2002) Would smokers with schizophrenia benefit from a more flexible approach to smoking treatment? *Addiction* 97: 785–793; discussion 795–800.

Millar, N.S. (2003) Assembly and subunit diversity of nicotinic acetylcholine receptors. *Biochem Soc Trans* 31: 869–874.

Moccia, F., Frost, C., Berra-Romani, R., Tanzi, F. and Adams, D.J. (2004) Expression and function of neuronal nicotinic ACh receptors in rat microvascular endothelial cells. *Am J Physiol Heart Circ Physiol* 286: H486–491.

Mongan, N.P., Baylis H.A., Adcock, C., Smith G.R., Sansom, M.S. and Sattelle, D.B. (1998) An extensive and diverse gene family of nicotinic acetylcholine receptor alpha subunits in *Caenorhabditis elegans*. *Receptors and Channels* 6: 213–228.

Mongan, N.P., Baylis, H.A., Adcock, C., Smith, G.R., Sansom, M.S. and Sattelle, D.B. (1998) An extensive and diverse gene family of nicotinic acetylcholine receptor alpha subunits in *Caenorhabditis elegans*. *Receptors and Channels* **6**: 213–228.

Mongan, N.P., Jones, A.K., Smith, G.R., Sansom, M.S. and Sattelle, D.B. (2002) Novel alpha7-like nicotinic acetylcholine receptor subunits in the nematode *Caenorhabditis elegans*. *Protein Sci* **11**: 1162–1171.

Nishiwaki, H., Nakagawa, Y., Kuwamura, M., Sato, K., Akamatsu, M., Matsuda, K., Komai, K. and Miyagawa, H. (2003) Correlations of the electrophysiological activity of neonicotinoids with their binding and insecticidal activities. *Pest Manag Sci* **59**: 1023–1030.

Patton, A., Knuth, S., Schaheen, B., Dang, H., Greenwald, I. and Fares, H. (2005) Endocytosis function of a ligand-gated ion channel homolog in *Caenorhabditis elegans*. *Curr Biol* **15**: 1045–1050.

Pereira, E.F., Hilmas, C., Santos, M.D., Alkondon, M., Maelicke, A. and Albuquerque, E.X. (2002) Unconventional ligands and modulators of nicotinic receptors. *J. Neurobiol* **53**: 479–500.

Phillips, H.A., Favre, I., Kirkpatrick, M., Zuberi, S.M., Goudie, D., Heron, S.E., Scheffer, I.E., Sutherland, G.R., Berkovic, S.F., Bertrand, D. and Mulley, J.C. (2001) CHRNB2 is the second acetylcholine receptor subunit associated with autosomal dominant nocturnal frontal lobe epilepsy. *Am J Hum Genet* **68**: 225–231.

Putrenko, I., Zakikhani, M. and Dent, J.A. (2005) A family of acetylcholine-gated chloride channel subunits in *Caenorhabditis elegans*. *J Biol Chem* **280**: 6392–6398.

Pym, L., Kemp, M., Raymond-Delpech, V., Buckingham, S., Boyd, C.A. and Sattelle, D. (2005) Subtype-specific actions of beta-amyloid peptides on recombinant human neuronal nicotinic acetylcholine receptors (alpha7, alpha4beta2, alpha3beta4) expressed in *Xenopus laevis* oocytes. *Br J Pharmacol* **146**: 964–971.

Rashid, M.H. and Ueda, H. (2002) Neuropathy-specific analgesic action of intrathecal nicotinic agonists and its spinal GABA-mediated mechanism. *Brain Res* **953**: 53–62.

Raymond-Delpech, V., Matsuda, K., Sattelle, B.M., Rauh, J.J. and Sattelle, D.B. (2005) Ion channels: molecular targets of neuroactive insecticides. *Invert Neurosci* **5**: 119–133.

Richmond, J.E. and Jorgensen, E.M. (1999) One GABA and two acetylcholine receptors function at the *C. elegans* neuromuscular junction. *Nat Neurosci* **2**: 791–797.

Robertson, A.P., Bjorn, H.E. and Martin, R.J. (1999) Resistance to levamisole resolved at the single-channel level. *Faseb J* **13**: 749–760.

Rogers, S.W., Andrews, P.I., Gahring, L.C., Whisenand, T., Cauley, K., Crain, B., Hughes, T.E., Heinemann, S.F. and McNamara, J.O. (1994) Autoantibodies to glutamate receptor GluR3 in Rasmussen's encephalitis. *Science* **265**: 648–651.

Sattelle, D.B. and Breer, H. (1990) Cholinergic nerve-terminals in the central-nervous-system of insects – molecular aspects of structure, function and regulation. *J Neuroendocrinology* **2**: 241–256.

Sattelle, D.B., Jones, A.K., Sattelle, B.M., Matsuda, K., Reenan, R. and Biggin, P.C. (2005) Edit, cut and paste in the nicotinic acetylcholine receptor gene family of *Drosophila melanogaster*. *Bioessays* **27**: 366–376.

Schoepfer, R., Conroy, W.G., Whiting, P., Gore, M. and Lindstrom, J. (1990) Brain alpha-bungarotoxin binding protein cDNAs and MAbs reveal subtypes of this branch of the ligand-gated ion channel gene superfamily. *Neuron* **5**: 35–48.

Seeburg, P.H. (2002) A-to-I editing: new and old sites, functions and speculations. *Neuron* **35**: 17–20.

Severance, E.G., Zhang, H., Cruz, Y., Pakhlevaniants, S., Hadley, S.H., Amin, J., Wecker, L., Reed, C. and Cuevas, J. (2004) The alpha7 nicotinic acetylcholine receptor subunit exists in two isoforms that contribute to functional ligand-gated ion channels. *Mol Pharmacol* **66**: 420–429.

Shaw, S., Bencherif, M. and Marrero, M.B. (2002) Janus kinase 2, an early target of alpha 7 nicotinic acetylcholine receptor-mediated neuroprotection against Abeta-(1-42) amyloid. *J Biol Chem* **277**: 44920–44924.

Shimomura, M., Okuda, H., Matsuda, K., Komai, K., Akamatsu, M. and Sattelle, D.B. (2002) Effects of mutations of a glutamine residue in loop D of the alpha7 nicotinic acetylcholine receptor on agonist profiles for neonicotinoid insecticides and related ligands. *Br J Pharmacol* **137**: 162–169.

Shimomura, M., Yokota, M., Matsuda, K., Sattelle, D.B. and Komai, K. (2004) Roles of loop C and the loop B-C interval of the nicotinic receptor alpha sub-unit in its selective interactions with imidacloprid in insects. *Neurosci Lett* **363**: 195–198.

Shimomura, M., Satoh, H., Yokota, M., Ihara, M., Matsuda, K. and Sattelle, D.B. (2005) Insect-vertebrate chimeric nicotinic acetylcholine receptors identify a region, loop B to the N-terminus of the Drosophila Dalpha2 subunit, which contributes to neonicotinoid sensitivity. *Neurosci Lett* **385**: 168–172.

Simmer, F., Moorman, C., van der Linden, A.M., Kuijk, E., van den Berghe, P.V., Kamath, R.S., Fraser, A.G., Ahringer, J. and Plasterk, R.H. (2003) Genome-wide RNAi of *C. elegans* using the hypersensitive *rrf-3* strain reveals novel gene functions. *PLoS Biol* **1**: E12.

Smit, A.B., Brejc, K., Syed, N. and Sixma, T.K. (2003) Structure and function of AChBP, homologue of the ligand-binding domain of the nicotinic acetylcholine receptor. *Ann N Y Acad Sci* **998**: 81–92.

Stetefeld, J. and Ruegg, M.A. (2005) Structural and functional diversity generated by alternative mRNA splicing. *Trends Biochem Sci* **30**: 515–521.

Tomizawa, M. and Casida, J.E. (2003) Selective toxicity of neonicotinoids attributa-ble to specificity of insect and mammalian nicotinic receptors. *Annu Rev Entomol* **48**: 339–364.

Tomizawa, M. and Casida, J.E. (2005) Neonicotinoid insecticide toxicology: mecha-nisms of selective action. *Annu Rev Pharmacol Toxicol* **45**: 247–268.

Touroutine, D., Fox, R.M., Von Stetina, S.E., Burdina, A., Miller, D.M., 3rd and Richmond, J.E. (2005) acr-16 encodes an essential subunit of the levamisole-resistant nicotinic receptor at the *Caenorhabditis elegans* neuromuscular junction. *J Biol Chem* **280**: 27013–27021.

Towers, P.R., Edwards, B., Richmond, J.E. and Sattelle, D.B. (2005) The *C. elegans* *lev-8* gene encodes a nicotinic acetylcholine receptor subunit (ACR-13) with roles in egg laying and pharyngeal pumping. *J Neurochem* **93**: 1–9.

Treinin, M. and Chalfie, M. (1995) A mutated acetylcholine receptor subunit causes neuronal degeneration in *C. elegans*. *Neuron* **14**: 871–877.

Treinin, M., Gillo, B., Liebman, L. and Chalfie, M. (1998) Two functionally dependent acetylcholine subunits are encoded in a single *Caenorhabditis elegans* operon. *Proc Natl Acad Sci USA* **95**: 15492–15495.

Ulloa, L. (2005) The vagus nerve and the nicotinic anti-inflammatory pathway. *Nat Rev Drug Discov* **4**: 673–684.

Unwin, N. (2005) Refined structure of the nicotinic acetylcholine receptor at 4A resolution. *J Mol Biol* **346**: 967–989.

Valiante, S., Capaldo, A., Virgilio, F., Sciarrillo, R., De Falco, M., Gay, F., Laforgia, V. and Varano, L. (2004) Distribution of alpha7 and alpha4 nicotinic acetylcholine receptor subunits in several tissues of *Triturus carnifex* (Amphibia, Urodela). *Tissue Cell* **36**: 391–398.

Watson, R., Jiang, Y., Bermudez, I., Houlihan, L., Clover, L., McKnight, K., Cross, J.H., Hart, I.K., Roubertie, A., Valmier, J., Hart, Y., Palace, J., Beeson, D., Vincent, A. and Lang, B. (2004) Absence of antibodies to glutamate receptor type 3 (GluR3) in Rasmussen encephalitis. *Neurology* **63**: 43–50.

Watson, R., Jepson, J.J., Bermudez, I., Alexander, S., Hart, Y., McKnight, K., Roubertie, A., Fecto, F., Valmier, J., Sattelle, D.B., Beeson, D., Vincent, A. and Lang, B. (2005) Alpha 7 acetylcholine receptor antibodies in two patients with Rasmussen encephalitis. *Neurology* **65**: 1802–1804.

Yassin, L., Gillo, B., Kahan, T., Halevi, S., Eshel, M. and Treinin, M. (2001) Characterization of the DEG-3/DES-2 receptor: a nicotinic acetylcholine receptor that mutates to cause neuronal degeneration. *Mol Cell Neurosci* **17**: 589–599.

Zhang, A., Kayser, H., Maienfisch, P. and Casida, J.E. (2000) Insect nicotinic acetylcholine receptor: conserved neonicotinoid specificity of [(3)H]imidacloprid binding site. *J Neurochem* **75**: 1294–1303.

5

Discovery of novel sodium channel inhibitors: a gene family-based approach

Jeff J. Clare

1 Introduction to voltage-gated sodium channels

Voltage-gated sodium (Na_V) channels are widely expressed in most electrically excitable cells. They activate in response to membrane depolarisation, and are responsible for rapid influx of Na^+ ions during the rising phase of an action potential. They also have a major influence on the resting potential of these cells and thus have a key role in regulating their excitability. The indispensable role of Na_V channels is well-illustrated by the lethal effects of a variety of neurotoxins which are highly potent and selective blockers of the channels (see Catterall, 1992). A well-known example of this is tetrodotoxin (TTX), the poisonous component of puffer fish, which can be deadly if not carefully removed before human consumption. The human genome encodes a family of ten different Na_V subtypes that share a high degree of sequence similarity (*Figure 1*) and which are highly conserved across mammalian species. Historically, these have often been classified into two major groups based on their sensitivity to TTX. The sensitive channels (TTX-s) include the major subtypes found in nervous tissue ($Na_V1.1$, 1.2, 1.3, 1.6 and 1.7) and muscle ($Na_V1.4$), whereas the TTX-resistant (TTX-r) channels include the cardiac subtype ($Na_V1.5$), plus two other subtypes ($Na_V1.8$ and 1.9) that have a highly restricted localisation in the peripheral neurons of the dorsal root ganglion (DRG). Since the latter are both primarily expressed in the nerves involved in pain signalling they are currently of major interest to the pharmaceutical industry as potential targets for improved analgesic drugs. In addition to the TTX-s and TTX-r channels there is a so-called 'atypical' subtype (Na_VX), which probably represents a distinct sub-family since it has not been demonstrated to function as a voltage-gated channel.

Molecular cloning of Na_V channels revealed them to be large and complex proteins (*Figure 2*). The pore-forming α subunit is organised into four homologous domains (I–IV) which each contain six transmembrane-spanning α-helical segments (S1–S6). Molecular studies of heterologously expressed channels have identified key regions of the α subunit that are involved in mediating the basic functions of the channel (for a review see Catterall, 2000). A re-entrant hydrophobic loop (SS1-SS2 or P-loop), located

Comparative Genomics and Proteomics in Drug Discovery, edited by John Parrington and Kevin Coward.
© 2007 Taylor and Francis Group.

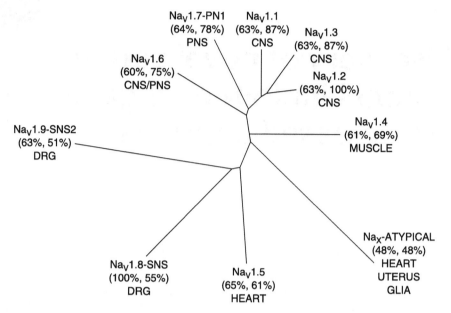

Figure 1. *Phylogenetic tree for human voltage-dependent sodium channels. The primary tissue(s) in which they are expressed and the percentage amino acid sequence identities relative to representative tetrodotoxin-sensitive ($Na_V1.2$, first value) and insensitive ($Na_V1.8$, second value) subtypes are indicated. (Reproduced with permission from Birkhauser-Verlag from Clare, 2005.)*

between the S5 and S6 segments in each domain, forms the channel pore and contains the determinants for ion selectivity. Within each domain there is a voltage sensor region located in the S4 transmembrane segments that contains stretches of positively charged amino acids occurring at every third position within the helix. During channel activation these charged amino acids have been shown to move as a result of electro-repulsive forces during membrane depolarisation, probably causing a conformational change that opens the channel pore. Channel inactivation is mediated by the cytoplasmic loop located between domains III and IV, which is thought to form a hydrophobic plug that blocks the channel by folding across the cytoplasmic end of the pore. Binding sites for the various toxin modulators of the channels have been identified (Catterall, 2000) and, in addition, key sites of interaction with Na_V blocking drugs have been identified on the inner face of the S6 segment from three out of the four domains (Ragsdale *et al.*, 1996; Yarov-Yarovoy *et al.*, 2001, 2002). Native Na_V channels also contain one or more β subunits of which there are four different subtypes, β1–4. These modulate channel function and have a role in trafficking the channel to the cell surface (Isom, 2001).

1.1 *Sodium channels and inherited diseases*

Na_V channels have been implicated in a number of human disease states. Seven different inherited disorders have been linked to mutations in Na_V genes, including two cardiac, three skeletal muscle and three neuronal syndromes (Lehmann-Horn and Jurkat-Rott, 1999; Head and Gardiner, 2003). To date, these Na_V 'channelopathies' have

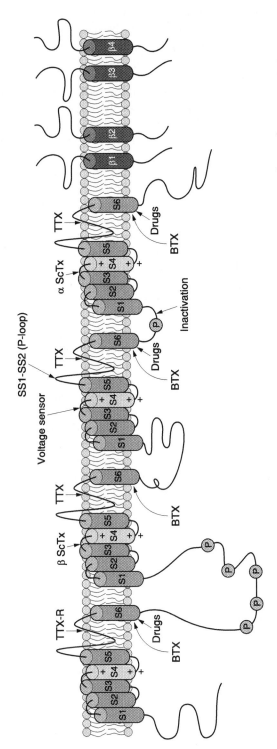

Figure 2. Structural and functional organisation of Na_V proteins. For clarity the voltage sensor, SS1-SS2, pore and ion selectivity regions are indicated only for domain II but are present in the other domains also. P in a circle indicates a demonstrated phosphorylation site, and the location of the key amino acid for TTX resistance is indicated in addition to the TTX binding determinants (for more details see text).

been found to affect four different α subunit subtypes and one β subunit subtype (see *Figure 3*). In general, these disorders are dominantly inherited (with the exception of motor endplate disease, MED) and their phenotypes are manifested episodically. That is, the mutations remain silent until triggered by particular physiological conditions, e.g. during exercise, cold temperatures, high serum [K$^+$], etc. Other potential triggers could conceivably include induction by drugs, though this has not yet been demonstrated. The episodic nature of these diseases probably reflects the rather subtle effects of these mutations on the properties of the channel. This is not unexpected since, given the fundamental role of Na$_V$ channels in electrical signalling, mutations causing more drastic effects are likely to cause embryonic lethality. As shown in *Figure 3*, a large number of different inherited mutations have now been identified and characterised. At first sight they appear to be randomly scattered throughout channel, though many are in or near regions involved in inactivation, including many of those associated with long QT syndrome (LQT3), potassium-aggravated myotonia (PAM), paramytonia congenita (PC) and generalised epilepsy with febrile seizures plus (GEFS+). Consistent with this, functional analysis of recombinant mutant channels confirms that a number of these do indeed cause defects in inactivation and lead to an increase in persistent Na$^+$ currents (INaP). These can be considered as 'gain-of-function' mutations and is consistent with their dominant mode of inheritance, e.g. LQT3, PAM, PC, hyperkalaemic periodic paralysis (HPP), GEFS+. In contrast, idiopathic ventricular fibrillation (IVF), severe myoclonic epilepsy of infancy (SMEI) and MED mutations either cause complete disruption of channel function, or reduced currents due to accelerated inactivation or changes in the voltage dependence of activation, and are thus 'loss-of-function' mutations. Interestingly, for mutations causing epilepsy, there appears to be a correlation between the nature of the mutation and the severity of the symptoms (Ceulemans *et al.*, 2004). Thus, truncating and mis-sense mutations in the pore-forming regions (S5–SS1–SS2–S6) of Na$_V$1.1 nearly always lead to a classic form of SMEI, whereas mis-sense mutations in the voltage sensor (S4) can lead to milder SMEI or GEFS+, and mis-sense mutations outside S4–S6 region mostly lead to GEFS+, or occasionally to milder forms of SMEI. At least three different GEFS+ mutations (R1648H, T875M, W1204R) cause 'gain-of-function' increases in the level of INaP, suggesting a highly plausible disease mechanism.

1.2 *Sodium channel blocking drugs*

Despite the fundamental physiological role of Na$_V$ channels it has been possible to develop therapeutically active Na$_V$ inhibitors that have relatively few side effects (for a review, see Clare *et al.*, 2000). However, these Na$_V$ blocking drugs were all discovered empirically, i.e. using traditional pharmacological methods, rather than by specifically targeting Na$_V$ channels, and only subsequently was it uncovered that they inhibited these channels. Thus, procaine was found to have local anaesthetic effects as early as 1905, but it was not till more than 50 years later that it was shown to have this effect by blocking Na$_V$ channels. Similarly, in 1936 procaine was found to correct certain cardiac arrhythmias when applied during heart surgery, but this was well before its effect on Na$_V$ channels had been discovered. Again, anticonvulsants like phenytoin (PTH), carbamazepine (CBZ) and lamotrigine (LTG) were all in clinical use before their action on Na$_V$ channels had been established.

Disease	Gene (channel)	Phenotype
Long QT syndrome 3 (LQT3)	SCN5A (Na$_V$1.5)	Ventricular arrhythmia, prolongation of cardiac action potential (and hence QT interval), can cause sudden death
Brugada syndrome (IVF, idiopathic ventricular fibrillation)	SCN5A (Na$_V$1.5)	Ventricular arrhythmia, ST interval elevated Can cause sudden death
Potassium aggravated myotonia (PAM)	SCN4A (Na$_V$1.4)	Myotonia, e.g. triggered by ingestion of potassium or, with delayed onset, after exercise
Paramyotonia congentia (PC)	SCN4A (Na$_V$1.4)	Myotonia during exercise, or exposure to cold
Hyperkalaemic periodic paralysis (HPP)	SCN4A (Na$_V$1.4)	Muscle weakness associated with high serum [K+]
Motor endplate disease (MED, jolting)	SCN8A (Mouse Na$_V$1.6)	MED–chronic ataxia, dystonia, lethal paralysis Jolting–cerebellar ataxia
Epilepsy: (GEFS)	SCN1A/SCN2A/SCN1B (Na$_V$1.1/Na$_V$1.2/β1)	Generalised epilepsy with febrile convulsions
(SMEI)	SCN1A (Na$_V$1.1)	Severe myoclonic epilepsy of infancy
(BFNIS)	SCN2A (Na$_V$1.2)	Benign familial neo-natal-infantile seizures

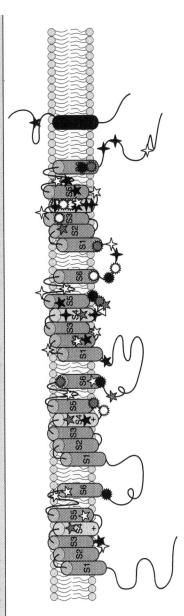

Figure 3. Voltage-gated sodium channelopathies. The different symbols indicate the location of mutations causing each of the inherited diseases associated with NaV channels, as indicated in the table.

Given the fundamental role of Na$_V$ channels, the question arises as to why these drugs do not cause major side effects. The availability of cloned Na$_V$ channels has provided tools to address this problem using a reductionist approach. That is, by studying Na$_V$ channels in a recombinant system they can be isolated from a neuronal environment in which many other ionic currents confound the interpretation of results. Such studies suggest there are three key properties of Na$_V$ blocking drugs that contribute to their therapeutic selectivity by favourably tipping the balance between the beneficial and detrimental *in vivo* effects of inhibiting Na$_V$ channels. These properties are illustrated in *Figure 4*, which shows data obtained using the cloned Na$_V$1.2 channel to study the mechanism of action of the anticonvulsant drug LTG. First, block by LTG is found to be around ten-fold more potent when the cells are held at a more depolarised potential (−60 mV vs. −90mV), indicating that the extent of block is voltage-dependent. This suggests that the drug acts more potently on neurons that are depolarised (for example, during seizure activity or pain signalling) than on those with normal resting potential. Second, the extent of block is use-dependent, that is, it progressively increases when cells are subjected to a series of short depolarising pulses. In the experiment shown in *Figure 4C* very little block is seen on the first pulse but inhibition accumulates with each successive pulse, reaching 25% block after 20 pulses. This suggests that neurons firing abnormally, i.e. with repetitive bursts of action potentials, such as occurs during pathological conditions, will be blocked to a greater extent than neurons firing normally. Third, the extent of use-dependent block is dependent on the frequency at which pulses are given and different inhibitors exhibit varying frequency dependence, probably as a result of differing kinetics of binding and dissociation. In this way it is possible, for example, for analgesic and anticonvulsant Na$_V$ blocking drugs to selectively inhibit neuronal populations that are firing at rapid 'pathological' rates, whilst having relatively little effect on slower-firing 'normal' neurons or indeed on slow-firing cardiac cells. This selective blockade of rapid neuronal firing patterns can be demonstrated in an *ex vivo* model of neuronal hyperexcitability, as shown in *Figure 4A*.

***

Figure 4. *Voltage and use-dependent action of Na$_V$ inhibitors. (A) Lamotrigine prevents repetitive firing of action potentials in an* in vitro *model. Synaptically evoked action potentials were measured by intracellular recordings from neurons in rat hippocampal slices superfused with normal artificial cerebrospinal fluid containing bicuculline (20 μM) but without Mg^{2+}. Sustained repetitive firing is observed which is blocked by lamotrigine (50 μM). (B) Potency of lamotrigine inhibition of hNa$_V$1.2 stably expressed in CHO cells is voltage dependent. Concentration-response curves were generated by measuring currents evoked by depolarising pulses to 0 mV from holding potentials (V$_h$) of either −90 mV (•) or −60 mV (▲) in the presence of different concentrations of lamotrigine. IC50 values were 641 μM and 56 μM at V$_h$ of −90 and −60 mV, respectively. (C) Inhibition of hNa$_V$1.2 by lamotrigine is use-dependent. Trains of depolarising pulses (20 ms duration, 10 Hz) from a Vh of −90 mV were applied and the currents elicited by each pulse normalised to the first pulse to remove the effects of tonic block. Inhibition by lamotrigine (100 mM, shaded circles) progressively increases with each additional pulse. (Reproduced with permission from Springer-Verlag from Xie et al., 1995 and 2001.)*

Another very interesting and useful property of Na_V inhibitors like LTG is that highly related structural congeners may exhibit a wide range of *in vivo* efficacy in different disease settings. Although first discovered as an anticonvulsant, LTG is also clinically effective in bipolar disorder and analgesia, and has been shown to be effective in a range of animal models including cerebral ischaemia, Parkinson's disease, nicotine and alcohol abuse, and anxiety (see Clare *et al.*, 2000). Several hundred structurally related

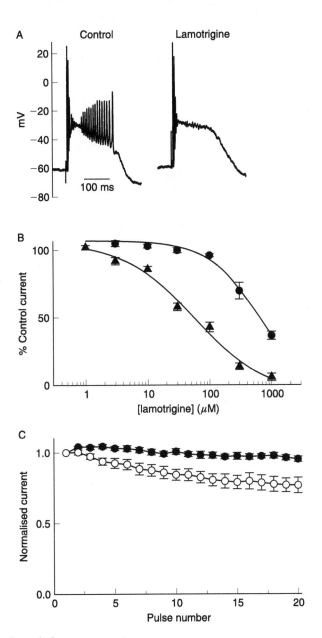

Figure 4. For legend please see opposite page.

compounds have been profiled in models of seizure and analgesia, identifying some that are therapeutically selective for either disease, and others with mixed activities that represent a spectrum in between these two extremes. Specific chemical alterations to the LTG template can be correlated with changes in therapeutic selectivity, though the molecular mechanisms behind these structure–activity relationships, for example at the level of interaction with the channel, are not yet understood. One intriguing possibility that has, in part, driven the gene family-based approach described here, is that this might reflect differing activity at the various Na_V subtypes.

2 A gene family-based approach to Na_V channels

The cloning of the first mammalian Na_V cDNAs precipitated numerous studies aimed at molecular characterisation of Na_V channels and, as discussed above, knowledge of their structure–function relationships is now extensive. However, information relating to brain channels has largely come from the study of the rat $Na_V1.2$ isoform which has been considered a prototypical brain subtype. Until recently, relatively few studies have characterised the other brain subtypes from rat and even less information has been available for the human orthologues. This partly reflects the technical challenges of handling Na_V channels and their cDNAs which makes them difficult to manipulate and heterologously express. However, owing to the importance of the human channels for drug discovery, and given the paucity of directly comparable data for the different subtypes, attention at GlaxoSmithKline has been focused on the human Na_V orthologues and we have taken a family-based approach by cloning, expressing and characterising almost the entire gene family (Chen *et al.*, 2000; Xie *et al.*, 2001; Burbidge *et al.*, 2002; John *et al.*, 2004). Our overall aims were several-fold: to define and compare their biophysical properties, to understand their individual functional roles *in vivo*, to determine their expression patterns and look for any changes associated with disease states, to identify any differences in these properties compared to the more widely studied rat orthologues, and to profile the activity of established Na_V blocking drugs, such as LTG and related compounds, in order to further understand their mechanism of action. This chapter will describe some of our results, illustrating these points mainly with examples relating to the four major brain subtypes, $Na_V1.1$, 1.2, 1.3 and 1.6.

2.1 *Distribution studies of human Na_V subtypes*

A gene family approach to the human Na_V subtypes provided early access to comprehensive sequence information for the entire human Na_V family and enabled the design and generation of a panel of highly subtype-specific oligonucleotide probes and anti-peptide antibodies. These tools were validated against the recombinantly expressed Na_V channel subtypes and then used to determine the pattern of Na_V mRNA and protein expression in human brain, both in normal tissue as well as in disease states (Whitaker *et al.*, 2000, 2001a, 2001b), allowing a comparison with the well-characterised distribution pattern of the rodent orthologues. Overall, the distribution patterns of the human subtypes were found to be similar to that reported for rodents in the regions studied (cerebellum, somato-motor cortex, hippocampus, basal ganglia and thalamus). However, a significant difference is that $Na_V1.3$, which based on rodent studies was previously considered to be an embryonic or neonatally expressed subtype, is found to be widely expressed in the human adult brain (*Figure 5*). This finding is consistent both in *in situ* hybridisation studies of mRNA and in immunological studies which demonstrate correspondingly wide expression of the human Na$_V$1.3 protein (*Figure 5*). In fact, subsequent immunological studies have now confirmed that this is also the case in rats (Lindia and Abbadie, 2003).

Having validated these subtype-specific oligonucleotide probes, we also used them to investigate potential changes in subtype mRNA expression in human disease states. As an example, *Figure 6* shows a comparison of the hybridisation patterns seen with Na$_V$1.2 and 1.3 probes in human epileptic hippocampus compared to controls. A clear reduction in Na$_V$1.2 mRNA can be seen in the epilepsy tissue compared to control, particularly in CA1, 2, 3 regions. To quantitate this, and also correct for neuronal loss that occurs in epilepsy patients, higher magnification images were developed and silver grains localised within individual cell bodies were counted within each neuronal population. The results show that in human epilepsy patients the level of Na$_V$1.2 mRNA is 20–30% reduced in CA1, 2, 3 regions of the hippocampus. Parallel data show that the level of 1.3 mRNA is increased by 45% in the CA4 (hilar) region of human epileptic hippocampus (*Figure 6A*). Although relatively subtle, these changes in mRNA abundance could have significant effects on level of Na$^+$ currents found in these cells, with correspondingly important changes in their excitability. Clearly, however, further work is required to determine if these changes play a causal role in epilepsy or are consequential to seizures or drug treatment.

In addition to confirming the overall mRNA distribution patterns obtained by *in situ* hybridisation, the human immunological studies have also confirmed and extended rodent studies of the sub-cellular localisation of the Na$_V$ subtypes in brain. The human Na$_V$1.2 subtype was shown to be uniquely concentrated along axons whereas the Na$_V$1.1, 1.3 and 1.6 were all found predominantly in neuronal cell bodies and proximal processes (*Figure 5*). This pattern has important implications for their respective functions in the brain and strongly suggests a specialised role for Na$_V$1.2 in action potential propagation and a role in modulating synaptic inputs and outputs for the other subtypes. This may represent an important species difference since in rodents, in addition to Na$_V$1.2, Na$_V$1.6 protein was also found to be present in unmyelinated axons within the brain (Caldwell *et al.*, 2000).

2.2 *Functional studies of human Na$_V$ subtypes*

A major impetus for cloning and expressing the entire family of human Na$_V$ subtypes was to compare their properties, since this has not previously been described. Even for the rat orthologues, this has not been comprehensively achieved and their properties can only be compared by combining data from different laboratories. Clearly, since the basic biophysical parameters are highly sensitive to minor changes in conditions and methodology this is a rather unsatisfactory comparison. In contrast, by generating a comprehensive set of data for each subtype in a single defined experimental system their properties can be more directly compared, and there is much greater confidence that any subtle differences are likely to be physiologically relevant. *Figure 7* summarises a comparison of some basic biophysical parameters for the four major human brain subtypes. As can be seen these have very similar properties and only

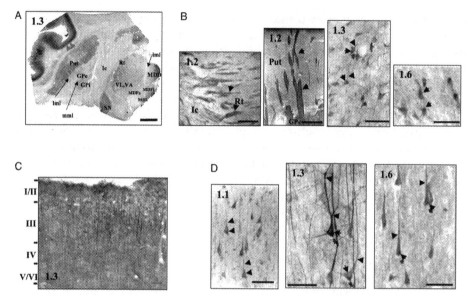

Figure 5. *Immunolocalisation of Na$_V$ subtypes in adult human brain tissue. (A) Na$_V$1.3 immunolocalisation in basal ganglia/thalamus (Ig, insular gyrus; Put, putamen; GPe, external globus pallidus; GPi, internal globus pallidus; lml; external medullary lamina of the globus pallidus, mml, medial medullary lamina of the globus pallidus; SN, substantia nigra; Ic, internal capsule; Rt, reticular thalamic nucleus; VL/VA, ventral thalamic nuclei; iml, internal medullary lamina of the thalamus; MDD, Mdfa, MDFi, MDV, medial dorsal thalamic nuclei; scale bar = 5 mm). (B) Magnified views of basal ganglia/thalamus showing differential subcellular localisation of Na$_V$1.2 (scale bars = 100 μM). Na$_V$1.2 (left hand panels, II) shows axonal staining with immunopositive fibre tracts in the external global pallidus (GPe - arrows) and in the thalamus, passing into the internal capsule (Ic) from the reticular thalamic nucleus (rt - arrows). In contrast, Na$_V$1.3 and Na$_V$1.6 (panels III and VI) show immunostaining in the soma of cells in the reticular thalamic nucleus (arrows) and, for Na$_V$1.3, also in the associated processes (arrowheads). (C) Na$_V$1.3 immunolocalisation in somato-motor cortex (roman numerals refer to the different cortical layers, scale bar = 200 μM). (D) Higher magnification of the somato-motor cortex, (scale bar represents 100 μM) showing Na$_V$1.1, 1.3 and 1.6 staining in the soma (arrows), proximal processes (arrowheads) and axon hillocks (asterisks) of cortical neurons. (Reproduced with permission by Elsevier from Whitaker et al. 2000.)*

very subtle differences can be observed. Interestingly, however, the most distinctive subtype in this system is Na$_V$1.2, which inactivates at more depolarised potentials ($V_{1/2inact}$ is 6–12 mV more positive) and recovers more rapidly from inactivation (τ_{inact} at the voltage giving maximum current is 2.6- to 3.4-fold less). While they appear to be rather minor, these differences could have important consequences *in vivo* and would be expected to lead to greater availability of Na$_V$1.2 channels than the other subtypes during neuronal depolarisations. Along with distribution studies that indicate Na$_V$1.2 channels have a unique axonal localisation within the human brain (*Figure 5*) this may reflect a more specialised functional role for Na$_V$1.2 in the propagation of action potentials in unmyelinated neurons in contrast with the other brain subtypes.

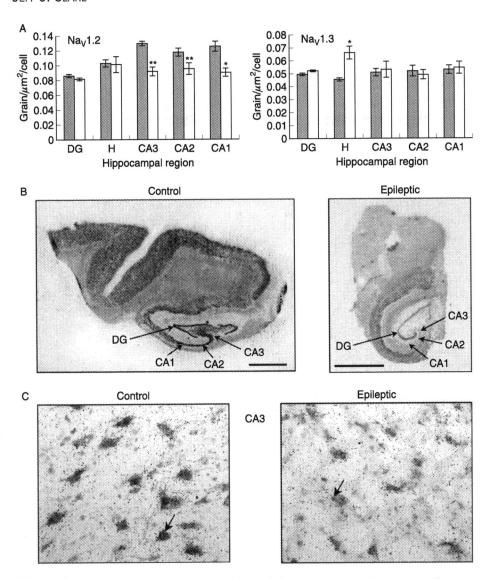

Figure 6. Changes in Na$_V$ channel mRNA levels in human epilepsy (A) Significant down-regulation of Na$_V$1.2 mRNA in regions CA1–3 is observed in hippocampus, whereas up-regulation of Na$_V$1.3 mRNA is observed in CA4 (hilus). (B) Macroscopic image of hippocampus from post-mortem control (left) and epileptic (right) human brain showing reduced expression of Na$_V$1.2 mRNA in CA1–3. (C) Higher magnification image showing cellular distribution of Na$_V$1.2 mRNA and reduced staining in the CA3 region of epileptic (right) human hippocampus compared to post-mortem control (left). (Reproduced with permission by Elsevier from Whitaker et al. 2001b.)

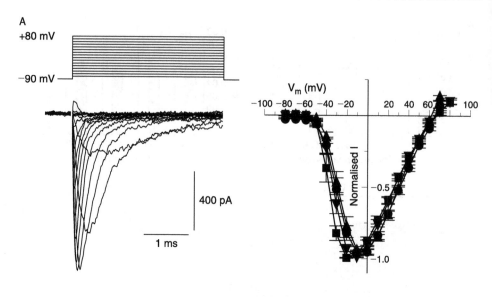

Figure 7. *Comparison of the biophysical properties of human brain Na$_V$ subtypes. (A) Left: representative traces showing depolarising pulses applied to HEK293 cells expressing human Na$_V$1.6 channels. Right: current-voltage relationships for the human brain Na$_V$ subtypes: Na$_V$1.1 (■), Na$_V$1.2 (•), Na$_V$1.3 (▲) and Na$_V$1.6 (▼). (B) Summary of the biophysical parameters measured: data are presented as mean ±SEM (n=4–7). (Reproduced with permission from Birkhauser-Verlag from Clare, 2005.)*

A more striking difference between Na$_V$1.2 and the other brain subtypes is in the level of INaP observed when expressed in a human cell background (*Figure 8*). Although the level of INaP observed varies from cell to cell and from clonal cell line to clonal cell line, in HEK293 cells the Na$_V$1.1, 1.3 and 1.6 subtypes consistently mediate greater levels of INaP than Na$_V$1.2 (typically 1–40% of total peak current for Na$_V$1.1, 1.3, 1.6, versus 0–5% for 1.2). The physiological significance of this finding remains to be determined, since these data are for recombinant channels expressed in a non-neuronal background. Nevertheless it is intriguing that the most widely studied brain Na$_V$ subtype (1.2), often considered the prototypical brain channel, is

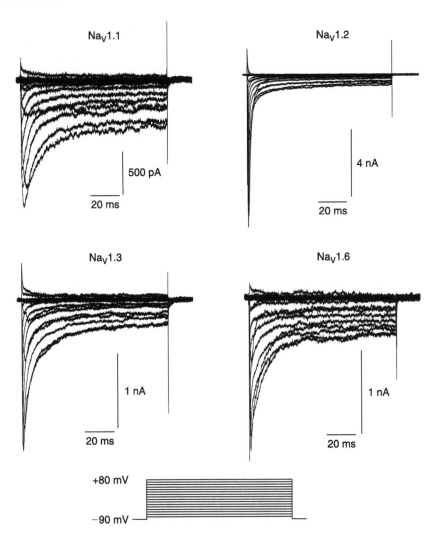

Figure 8. *Currents mediated by human Na$_V$1.1, 1.3 and 1.6 subtypes show a prominent persistent component when expressed in a human cell background. Representative inward currents evoked by a series of depolarising pulses (100 ms duration, see bottom) are shown for all four brain subtypes when stably expressed in HEK293 cells. Currents decay with biphasic kinetics consisting of a rapid component (transient) and a sustained component (persistent). The sustained component persists for at least 100 ms and, with the exception of Na$_V$1.2, can comprise a large proportion of the total current (up to 40% in some cells, depending on conditions used). (Reproduced with permission from Birkhauser-Verlag from Clare, 2005.)*

atypical in its propensity to mediate persistent current as well as having more subtle differences in its other biophysical properties.

The INaP observed in this system is probably modulated via trimeric G-proteins, since the level of INaP is found to decay when CsF is present in the recording pipette solution (Burbidge *et al.*, 2002) and this is known to indirectly modulate G-protein activation (Chen and Pennington, 2000). G-protein involvement is consistent with other studies that show co-expression of G$\beta\gamma$ subunits in HEK293 cells induces increased levels of INaP mediated by rat Na$_V$1.2 (Ma *et al.*, 1997) and, more recently, by human Na$_V$1.1 (Mantegazza *et al.*, 2005). The finding that the different brain Na$_V$ subtypes have differing intrinsic basal levels of INaP could have profound implications for their roles *in vivo*, both in normal and in pathological settings. A higher level of intrinsic INaP mediated by Na$_V$1.1, 1.3 and 1.6, together with their somato-dendritic localisation within brain neurons (*Figure 5*), is consistent with these subtypes having a major influence on membrane potential in the cell body, as well as on processing of synaptic inputs, on controlling frequency of firing and on shaping burst firing behaviour, for example, during epileptiform hyperexcitability.

2.3 *Pharmacological studies of the human Na$_V$ subtypes*

Cloning and expression of the human Na$_V$ subtypes has enabled the selectivity of existing therapeutic inhibitors to be profiled. It is conceivable that this analysis could reveal differences in potency at the Na$_V$ subtypes that might give insight into the variation in relative efficacy of the various Na$_V$ inhibitors in different diseases and disease models. For example, as described above, this may partly explain the wide spectrum of *in vivo* efficacy in models of pain and seizure that can be observed in a series of structural analogues derived from lamotrigine. However, to date little intrinsic subtype selectivity within the Na$_V$ family has been found with the commonly used Na$_V$ inhibitors. For example, lamotrigine is found to have similar potency for voltage-dependent tonic block at each of the major brain subtypes (see Clare, 2005). Similarly, commonly used Na$_V$ inhibitors like lamotrigine appear to show relatively little selectivity between blockade of INaP and of transient currents (Chao and Alzheimer, 1995; Segal and Douglas, 1997; Taverna *et al.*, 1998; Spadoni *et al.*, 2002; Clare, 2005).

Detailed profiling of compounds against the Na$_V$ family of subtypes using conventional patch clamp electrophysiology is a highly labour-intensive undertaking. However, recent technical advances such as the development of planar array electrophysiology instruments (Wood *et al.*, 2004) have dramatically increased the throughput of this analysis and opened up the possibility of profiling and comparing large numbers of compounds. Importantly, it is possible to measure more complex properties such as use-dependence in a relatively high-throughput manner using these instruments (Shroeder *et al.*, 2003). As an illustration of this, *Figure 9* shows a comparison of the potency of both tonic and use-dependent block of several different commonly used Na$_V$ inhibitors (representative of drugs with analgesic, anti-convulsant and local anaesthetic activity) at three different Na$_V$ subtypes, using data obtained with an Ionworks[HT] instrument (Tim Dale, personal communication). The various compounds show subtly different profiles of activity but none are more than 10-fold selective for any of the subtypes measured. Nevertheless, this technology holds promise for the future as it allows a vast increase in the number of compounds that can be screened and should enable the discovery of novel compounds with greater levels of selectivity. The *in vivo* activity of such molecules remains to be determined, though it can be speculated that this may lead to improved levels of efficacy and/or therapeutic selectivity.

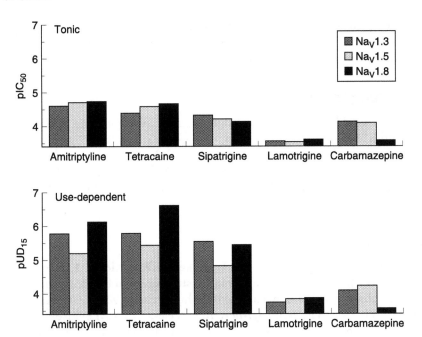

Figure 9. Comparison of the potency of tonic and use-dependent block of several commonly used Na$_V$ inhibitors for three different Na$_V$ subtypes using planar array electrophysiology (Ionworks^{HT}). Voltage-clamp recordings were made from cells stably expressing human Na$_V$1.3, Na$_V$1.5 or Na$_V$1.8 subtypes. Trains of 20 ms depolarising pulses from holding potentials of –90 mV (Na$_V$1.3 and Na$_V$1.8) or –105 mV (Na$_V$1.5) were applied at a frequency of 10 Hz (Na$_V$1.3 and 1.8) or 3.5 Hz (Na$_V$1.5). Test compound was then added (to four cells for each concentration) and a second set of depolarising pulses applied, exactly as described above. Tonic inhibition was determined by plotting the current elicited by the first pulse in the train as a percentage of pre-compound current. The upper histogram depicts the pIC50 values of tonic block calculated from the concentration response curves. Use-dependent inhibition was assessed by calculating the ratio for the current evoked by the tenth/first pulse in the absence and presence of compound. The lower histogram illustrates pUD15 values (the concentration of inhibitor giving 15% use-dependent block) calculated from the resultant concentration response curve.

3 Summary and future prospects

Advances in assay and screening technologies in recent years have dramatically increased the tractability of Na$_V$ channels for drug discovery using molecular target-driven approaches. In parallel with this, advances in our understanding of Na$_V$ channel structure–function relationships, the Na$_V$ drug binding site and the advent of the first three-dimensional crystal structure for a mammalian voltage-gated ion channel (Long *et al.*, 2005) have provided potential opportunities to include structural insights into the development of Na$_V$ modulators in the future. Together these developments have re-invigorated Na$_V$ drug discovery within the pharmaceutical industry. It is hoped that the gene family approach described here will contribute to these advances by providing a variety of tools and reagents that can be used to facilitate the development of a new generation of Na_V modulators and to further characterise the biology of these important drug targets.

Acknowledgements

The author would like to acknowledge friends, colleagues and collaborators at GlaxoSmithKline and elsewhere who, though too numerous to mention, have supported and contributed to Na_V research at GSK. Particular thanks go to Mike Romanos for his support and to Steve Burbidge, Yuhua Chen, Tim Dale, Matt Hall, Del Trezise, Andy Powell, Will Whitaker and Xinmin Xie for direct contributions to the data shown in the figures.

References

Burbidge, S.A., Dale, T.J., Powell, A.J., Whitaker, W.R.J., Xie, X.M., Romanos, M.A. and Clare, J.J. (2002) Molecular cloning, distribution and functional analysis of the Na_V1.6 voltage-gated sodium channel from human brain. *Mol Brain Res* 103: 80–90.

Caldwell, J.H., Schaller, K.L., Lasher, R.S., Peles, E. and Levinson S.R. (2000) Sodium channel Na_V1.6 is localised at nodes of Ranvier, dendrites and synapses. *Proc Natl Acad Sci* 97: 5616–5620.

Catterall, W.A. (1992) Cellular and molecular biology of voltage-gated sodium channels. *Physiol Rev* 72: S15–48.

Catterall, W.A. (2000) From ionic currents to molecular mechanisms: the structure and function of voltage-gated sodium channels. *Neuron* 26: 13–25.

Ceulemans, B.P., Claes, L.R. and Lagae, L.G. (2004) Clinical correlations of mutations in the SCN1A gene: from febrile seizures to severe myoclonic epilepsy in infancy. *Ped Neurol* 30: 236–243.

Chao, T.I. and Alzheimer, C. (1995) Effects of phenytoin on the persistent Na+ current of mammalian CNS neurones. *Neuroreport* 6: 1778–1780.

Chen, Y. and Pennington, N.J. (2000) Competition between internal AlF_4^- and receptor mediated stimulation of dorsal raphe neuron G-proteins coupled to calcium current inhibition. *J Neurophysiol* 83: 1273–1282.

Chen, Y.H., Dale, T.J., Romanos, M.A., Whitaker, W.R.J., Xie, X.M. and Clare, J.J. (2000) Cloning, distribution and functional analysis of the type III sodium channel from human brain. *Eur J Neurosci* 12: 4281–4289.

Clare, J.J. (2005) Current approaches for the discovery of novel NaV channel inhibitors for the treatment of brain disorders. In: Coward, K. and Baker, M.D. (eds) *Sodium Channels, Pain, and Analgesia.* Birkhauser Verlag, Basel: 23–62.

Clare, J.J., Tate, S.N., Nobbs, M. and Romanos, M.A. (2000) Voltage-gated sodium channels as therapeutic targets. *Drug Discovery Today* 5: 506–520.

Head, C. and Gardiner, M. (2003) Paroxysms of excitement: sodium channel dysfunction in heart and brain. *BioEssays* 25: 981–993.

Isom, L.L. (2001) Sodium channel beta subunits: anything but auxiliary. *Neuroscientist* 7: 42–54.

John, V.H., Main, M.J., Powell, A.J., Gladwell, Z.M., Hick, C., Sidhu, H.S., Clare, J.J., Tate, S. and Trezise, D.J. (2004) Heterologus expression and functional analysis of rat $Na_V1.8$ (SNS) voltage-gated sodium channels in the dorsal root ganglion neuroblastoma cell line ND7-23. *Neuropharmacol* **46**: 425–438.

Lehmann-Horn, F. and Jurkat-Rott, K. (1999) Voltage-gated ion channels and hereditary disease. *Physiol Rev* **79**: 1317–1372.

Lindia, J.A. and Abbadie, C. (2003) Distribution of the voltage-gated sodium channel $Na_V1.3$-like immunoreactivity in the adult rat central nervous system. *Brain Res* **960**: 132–141.

Long, S.B., Cambell, E.B. and MacKinnon, R. (2005) Crystal structure of a mammalian voltage-dependent shaker family K+ channel. *Science* **309:** 897–903.

Ma, J.Y., Catterall, W.A. and Scheuer, T. (1997) Persistent sodium currents through brain sodium channels induced by G-protein βγ subunits. *Neuron* **19**: 443–452.

Mantegazza, M., Yu, F.H., Powell, A.J., Clare, J.J., Catterall, W.A. and Scheuer, T. (2005) Molecular determinants for modulation of persistent sodium current by G-protein βγ subunits. *J. Neurosci* **25**: 3341–3349.

Ragsdale. D.S., McPhee, J.C., Scheuer, T. and Catterall, W.A. (1996) Common molecular determinants of local anesthetic, antiarrhythmic and anticonvulsant block of voltage-gated Na^+ channels. *Proc Natl Acad Sci* **93**: 9270–9275.

Schroeder, K., Neagle, B., Trezise, D.J. and Worley, J. (2003) IonWorks: a new high-throughput electrophysiology measurement platform. *J Biomol Screening* **8**: 50–64.

Segal, M.M. and Douglas, A.F. (1997) Late sodium channel openings underlying epileptiform activity are preferentially diminished by the anticonvulsant phenytoin. *J Neurophysiol* **77**: 3021–3034.

Spadoni, F., Hainsworth, A.H., Mercuri, N.B., Caputi, L., Martella, G., Lavaroni, F., Bernardi, G. and Stefani, A. (2002) Lamotrigine derivatives and riluzole inhibit INa,P in cortical neurons. *Neuroreport* **13**: 1167–1170.

Taverna, S., Mantegazza, M., Franceschetti, S. and Avanzini, G. (1998) Valproate selectively reduces the persistent fraction of Na^+ current in neocortical neurons. *Epilepsy Res* **32**: 304–308.

Whitaker, W.R.J., Clare, J.J., Powell, A.J., Chen, Y., Faull, R.L.M. and Emson, P.C. (2000) Distribution of voltage-gated sodium channel α and β subunit mRNAs in human cerebellum, cortex and hippocampal formation. *J Comp Neurol* **422**: 123–139.

Whitaker, W.R.J., Faull, R.L.M., Waldvogel, H., Plumpton, C.J., Emson, P.C. and Clare, J.J. (2001a) Comparative distribution of voltage-gated sodium channel proteins in human brain *Mol Brain Res* **88**: 37–53.

Whitaker, W.R.J., Faull, R.L.M., Emson, P.C. and Clare, J.J. (2001b) Changes in mRNAs encoding voltage-gated sodium channel types II and III in human epileptic hippocampus. *Neurosci* **106**: 275–285.

Wood, C., Williams, C. and Waldron, G.J. (2004) Patch-clamping by numbers. *Drug Discovery Today* **9**: 434–441.

Xie, X., Lancaster, B., Peakman, T. and Garthwaite, J. (1995) Interaction of the antiepileptic drug lamotrigine with recombinant rat brain type IIA Na^+ channels and with native Na^+ channels in rat hippocampal neurones. *Pflugers Arch* **430**: 437–446.

Xie, X.M., Dale, T.J., John, V.H., Cater, H.L., Peakman, T.C. and Clare, J.J. (2001) Electrophysiological and pharmacological properties of the human brain type IIA Na^+ channel expressed in a stable mammalian cell line. *Pflug Arch* **441**: 425–433.

Yarov-Yarovoy, V., Brown, J., Sharp, E., Clare, J.J., Scheuer, T. and Catterall, W.A. (2001) Molecular determinants of voltage-dependent gating and binding of pore-blocking drugs in transmembrane segment IIIS6 of the sodium channel α subunit. *J Biol Chem* **276**: 20–27.

Yarov-Yarovoy, V., McPhee, J.C., Idsvoog, D., Pate, C., Scheuer, T. and Catterall, W.A. (2002) Role of amino acid residues in transmembrane segments IS6 and IIS6 of the Na$^+$ channel α subunit in voltage-dependent gating and drug block. *J Biol Chem* **38**: 35393–35401.

6

'Omics' in translational medicine: are they lost in translation?

John A. Bilello

Advances in 'omics'[1] technologies (genomics, transcriptomics, proteomics and metabonomics) were touted as having the potential to revolutionize our approach to disease diagnosis, prognostication and development of novel therapeutics. However, the promise of rapid advances in medicine 'from the lab bench to the bedside' has not manifested as of yet. Indeed it appears that the translational applications of genomic-based research have preceded the development of (*i*) a conceptual framework for disease understanding, (*ii*) effective tools that can exploit the vast amounts of data derived from these efforts and (*iii*) a process to systematically move the products of basic research towards standardized, reproducible, clinical diagnostic tools. This chapter will focus upon the translation of 'omic' information into disease understanding and effective therapeutic management.

1 Introduction

The application of platform technologies which enhance understanding of the disease processes permit us to identify therapeutic points of intervention (tractable targets). Strategies built upon 'omic' technologies should also provide mechanistic insight leading to the identification of biomarkers or panels that would be useful in diagnosis, patient stratification and disease state monitoring; all aspects key to markedly enhanced patient care. In addition to providing mechanistic insight into disease, therapeutic targets and/or drug efficacy, 'omic' technologies may be able to distinguish adverse effects from exposure to drug or toxicant and thus provide improved prediction of human outcomes.

However despite the ongoing investment in new technologies, there is considerable evidence that these changes have not delivered a new era of prevention or treatment of disease which was promised by biomedical research. Indeed the number of new drugs (new chemical entities, NCEs) or biologics submissions to the US Food and Drugs

---

[1] The 'omics' suffix has come to signify the measurement of the entire complement of a given category of biological molecules and information.

Comparative Genomics and Proteomics in Drug Discovery, edited by John Parrington and Kevin Coward.
© 2007 Taylor and Francis Group.

Administration (FDA) and other regulatory agencies worldwide has decreased significantly over the past decade. Perhaps in parallel the number of new protein diagnostics has markedly decreased over the same time period (Anderson and Anderson, 2002). In contrast, current costs of taking an NCE to market has increased to nearly $1 billion (Tufts Center for Drug Development, 2001; Rawlins, 2004) which represents a barrier to investment in innovative therapeutics particularly those with limited market value or for diseases which effect third world or economically challenged populations.

2 'Omic' technologies: a capsule

'Omic' platforms, while not necessarily at the scale of describing the totality of the genome, proteome or metabalome, are sensitive technologies for detection, quantitation and identification of DNA, mRNA, proteins and metabolites derived from complex body tissue and fluids. In a recent review on 'omic' technologies in drug development (Bilello, 2005) I detailed aspects of the 'omic' revolution, focusing on both the progress and the shortcomings. In a relatively short timespan there have been a number of recent reviews focusing on various aspects of individual platform technologies and I refer readers to these (Morel *et al.*, 2004; Thadikkaran *et al.*, 2005; David *et al.*, 2005; Giallourakis *et al.*, 2005; Lindon *et al.*, 2004; Fan and Hedge, 2005; Watkins, 2004). If we were to bottom line 'omics': Gen*omics* platforms catalogue and measure genes. Recent advances in high-throughput genotyping technologies have realized the possibility of performing large-scale, high-resolution genetic studies in human complex diseases. As an outgrowth of these technologies, single nucleotide polymorphisms (SNPs) are the markers of choice for polygenic diseases due to their frequent occurrence and simple mutational dynamics. Transcript*omic* platforms characterize and quantify gene expression at the level of messenger RNA (mRNA). Prote*omic* platforms determine the identities and concentrations of protein analytes within a biological tissue or fluid. Proteomics is a read-out of protein expression at a defined time, under a specific set of conditions or disease states which provide the basis for comparison. In contrast functional proteomics determines protein–protein, protein–DNA, and protein–RNA interactions in an attempt to identify functionality. Metabol*omics* (also known as metabon*omics*) platforms characterize and quantify small molecules within a biological tissue or fluid.

3 Defining translational research/medicine

In medical research, the terms 'translational medicine' and 'translational research' while widely used have a variety of definitions and interpretations (Webb and Pass, 2004; Horig *et al.*, 2005). To many, the term *translational* represents the interface between the research laboratory ('bench') and patient care ('bedside'). This process is often looked upon as the transitioning of a basic research finding to the clinic-based treatment, prevention or diagnosis of a specific disease. In my own view, translational medicine is a spectrum of activities embracing basic research, disease understanding (including epidemiology, behavioural studies), bridging studies and the incorporation of new therapies or technological innovations into the practice of medicine.

    Translational research is focused primarily upon enabling studies. These can be categorized as human and non-human studies which bridge preclinical and clinical research.

In a limited view such studies are non-clinical studies intent upon incorporating new paradigms, therapies or technological innovations into patient-oriented research. Such studies may: (*i*) evaluate novel diagnostic methodologies, (*ii*) assess new technologies, (*iii*) characterize disease phenotypes (imaging, patient stratification) and (*iv*) identify biological markers (biomarkers, surrogate markers). Given that view, translational research provides a series of tools that define the biological effects of disease pathogenesis and therapeutics in humans. It also serves as an incubator for a new generation of clinician scientists who are at home at the lab bench, computer and the bedside.

So what distinguishes 'translational' from 'academic' medicine? Academic medicine is often likened to a three-legged stool comprising scientifically based medical education, patient care and research. Is translational medicine a two-legged stool comprised of patient care and research? Or is it only one leg: research? Personally I feel that translational medicine is modern medical science embracing research (basic and 'applied'), while being at the core of advancing contemporary medical education and the delivery of state of the art medical care. It is an all-inclusive discipline, spanning institutes of biotechnology and biomedicine, pharma, academic centres and teaching hospitals. In the world of academic medicine, translational research is poised at the threshold of becoming the engine of a massive improvement in predictive, preventive and personalized medicine. In the pharmaceutical industry, translational research lies between preclinical research in cellular and animal model systems and proof of concept and dose selection in humans. Translational research is at the precipice of enabling better decision making at the preclinical–clinical interface, e.g. by using biomarkers to support the choice of stronger therapeutic candidates.

4 Are 'omics'-derived markers being 'Lost in Translation'?[2]

'Omic' technologies are fuelling the quest for improved disease understanding and the management of disease. High-throughput screening of DNA variants, transcripts, proteomic and metabolomic profiles can identify determinants of susceptibility and markers which predict disease onset, progression and response to therapy. One way to bring this into sharper focus is to remind the reader that the end products of 'omic' technologies, e.g. genes or proteins, may directly lead to the development of clinically useful diagnostic tests and biomarkers to aid in the development of new therapeutics.

The majority of genetic tests which are used in clinical practice are those involved in the diagnosis of rare, single-gene disorders in a limited proportion of the population. In addition there are a small number of genetic tests used for population-based newborn screening and carrier testing in ethnic groups at high risk for selected diseases. The basis for the current limited application of genetic-based tests to clinical medicine may be that common complex diseases result from interactions between

---

[2] One of my favourite websites is 'Lost in Translation' (http://www.tashian.com/multibabel/). This 'time waster' is a site that will take English text and translate it into several languages and then back to English. The sites translation software 'is almost good enough to turn grammatically correct, slang-free text from one language into grammatically incorrect, barely readable approximations in another. . . .' In addition there is this meandering film of the same name where the characters are anchored in their surroundings, unable to get in the flow of a different culture. In a way that is where I feel we are with omics and translational medicine.

many low-penetrant genes and environmental factors that limit the ability to test individuals for genetic susceptibility and to tailor interventions. In their review of the utilization of genetic tests in public health settings, Yoon *et al.* (2001) suggest that only a small percentage, perhaps 10%, have had a significant impact.

Clearly pharmacogenomics (the study of the inherited nature of differences in drug disposition and effects) offers the potential of using DNA-based diagnostic tests to develop therapeutic products that are efficacious and safe for specific populations. The caveat is that patient response to treatment is due in part to genetic makeup and in part to other factors, such as their diet, age, gender, weight and concurrent medications. Such variability has hindered developing genetic tests to the level where they are effectively used clinically. By way of example, mutations in two genes, BRCA1 and BRCA2, are associated with predisposition for inherited breast and ovarian cancer and are identified in 5–10% of women with breast or ovarian cancer. Genetic testing for these mutations has been available clinically since the late 1990s; however, population-based screening is not recommended because of the complexity of test interpretation and limited data on clinical validity and utility (US Preventative Services Taskforce, 2005). As a harbinger of things to come, testing for HER-2 expression using pharmacogenomic tools is an example of targeting a therapy to a specific genetic profile (Pegram and Slamon, 2000). While the potential number of patients who will respond to Herceptin is limited, there are two FDA-approved tests. One is a fluorescence *in situ* hybridization (FISH) assay which measures gene amplification, while the other measures serum HER-2 protein levels to determine HER-2 status and select patients for treatment. However, it is disappointing to note that FDA approval came ten years after Slamon and his co-workers demonstrated that amplification of the HER-2/neu gene was a significant predictor of both overall survival and time to relapse in patients with breast cancer (Slamon *et al.*, 1987). A major reason for this delay was the fact that there was a marginal need to progress the development of a diagnostic test for HER-2/neu expression until Herceptin was clearly on its way to approval.

Similar problems have occurred with protein-based diagnostics. In their review, Anderson and Anderson (2002) have pointed out that the rate of development of FDA-approved diagnostic tests for proteins in plasma has fallen. Since 1994 about a dozen new protein analytes have been added, despite massive discovery efforts in plasma proteomics.

The process of translation of a discovery in the laboratory (e.g. of a potential biological marker) to its clinical implementation seems to be where the products of 'omic' platforms get lost.

5 Is it a question of getting the biology straight?

Monya Baker suggests 'It's the biology, stupid' in discussing that what appears to be a lack of understanding of pathophysiology may be an obstacle to transitioning potential markers (Baker, 2005). There are too many examples of biomarkers which have proved to be useful in a model system but become essentially useless in human patient populations. One aspect of the biology problem is that most of the discovery has focused upon differential gene expression in two extremes, e.g. differentiated from undifferentiated cell types, diseased states from controls. The difficulty is that such an approach does not necessarily reflect the true biology. Often the markers identified for the disease state are those reflective of pathologic changes in the tissues and may not be reflective of a disease-specific process. Since most disease states have aspects of inflammation, and tissue remodelling, biomarkers related to these events may be elevated in serum or lymphocytes from diseased subjects. By way of example, increased C-reactive protein (CRP) serum levels are related to smoking, diseases with a chronic inflammatory component (Rifai and Ridker, 2001; Visser *et al.*, 1999; Man and Sin, 2005), cardiovascular disease, obesity and oestrogen status. As a corollary, a variety of medications such as aspirin, statins and thiazolidinediones tend to decrease CRP. If one finds an increase in CRP in patients with chronic obstructive pulmonary disease (COPD) as compared to control, that does not represent a disease-specific biomarker for COPD nor does it suggest that aspirin is a disease-modifying drug for COPD patients. While the CRP analogy is particularly easy to comprehend, there are a large number of mass spectroscopy-based approaches that have identified the same series of alpha-2 macroglobulins, haptoglobins and complement components linked to multiple diseases (Baker, 2005; Bilello, unpublished). All too rarely has the experimental design of disease-related discovery efforts focused upon related diseases, diseases with similar sequelae or longitudinal aspects of disease manifestation. Perversely, multivariate analysis performed on a data set with a large number of variables has yielded statistically significant patterns of analytes capable of segregating two groups based strictly upon chance. Disease understanding and appropriate experimental design can segregate markers which are both statistically and biologically significant.

Equally as important as understanding the biology is the understanding that human biology is extraordinarily complex. Human populations are diverse: at a genetic level they are out-bred and polymorphic. Human subjects are exposed to a wide variety of external factors which contribute further to the heterogeneity of the population. Disease itself can result from a simple or single genetic aberration, e.g. cystic fibrosis; however, there are many more diseases that have a complex pattern of inheritance, such as in the cases of diabetes, asthma, cancer and mental illness. For complex diseases, more than one mutation is required to initiate the disease and/or multiple genes may each contribute to individuals' susceptibility to a disease and how someone responds to environmental triggers and factors.

6 What are the factors contributing to the inability to rapidly move the products of omic platforms towards standardized, reproducible, clinical diagnostic tools?

At present we are at a stage where major barriers interfere with the progress in evaluating potentially important discoveries and translating them into clinical tests which can be used in routine clinical settings. The barriers to the effective translation of 'omic'-based research to the clinic appear to revolve around validation, in particular biological as opposed to analytical validation. In analytical laboratories variability of the results is considered on an individual basis, and accuracy (trueness and precision) refers to individual analytical results. Trueness refers to how close to the true value the average of the results is, precision to the scatter of the results around their average, while accuracy is the combination of the two characteristics. In developing assays for research or clinical settings, the analytical validity or accuracy can be determined by the use of the appropriate controls or standard reference materials (SRM) when available. Essentially analytical validation establishes that the assay accurately measures what it is designed to measure while establishing what factors are involved in maintaining the robustness of the assay. Quality assurance, in terms of intra- and interassay of precision and accuracy must always be performed to check the overall analytical work. This can be conveniently performed if appropriate SRMs with known concentrations of the analyte(s) under study are available.

In terms of the biological or perhaps more appropriately clinical validation, the confounding issues are of even greater proportions. The following discussion of biological/clinical validation focuses on biomarkers. Biomarker is a much-abused term that includes biological markers, surrogates, prognostics and diagnostics (Morel et al., 2004). This is actually a relative hierarchy based both upon degree of statistical validation with human disease and acceptance in the external medical and/or regulatory community. A biologic marker is a test or set of tests which correlates with alterations in a given physiologic or biochemical pathway. Although the biologic marker may fluctuate in human disease it does not have a known or as yet validated association with alterations in disease outcome measurements. Thus in the hierarchy indicated above, a biological marker is a low stringency biomarker. Were the variation in the biological marker to be validated as a predictor of clinical (human) outcome it could thus become a surrogate. If the biological marker is validated as a predictor of the likely progression of a specific disease it is a prognostic. A prognostic's indication of the progression of the specific disease may be in reference to the unmodified pathology or in reference to the likelihood of a patient to respond to therapy. The highest degree of both statistical validity and acceptance in medical practice is that held by a diagnostic.

Another major obstacle to the development of potential biomarkers from omic discovery platforms is in the quality of the sample sets used in both the discovery and validation processes. Considering the variability in human systems, it is not difficult to comprehend that one will need to evaluate a relatively large number of well-characterized clinical specimens. Given the fact that most human diseases can exist as a spectrum of severity, all too often clinical samples are not sufficiently stratified. Imaging technologies such as MRI or CT scans, which can be used to stage patients, are expensive and are rarely applied within the discovery phase of development. Sample acquisition and preparation methods are often not standardized and storage conditions may be variable and not optimal with respect to both temperature and length of time stored. Lack in sufficient detail in the sample protocol leads to short-term variability (hour to hour, day to day) in biological characteristics due to, for example, diurnal variation, time since last meal, or whether the subject is sitting vs. lying down, etc. In metabolomics, in particular, within-patient medium-term variability (month to month) can arise due to seasonal changes in diet or exercise. Lastly long-term change (year to year) may result from changes in disease severity, co-morbidities, medications, institutionalization or purposeful dietary changes over time. Unlike clinical trials, which collect samples under stringently defined and documented protocols, sample sets used in discovery may have limited documentation and may not be balanced in terms of age, sex, and disease severity.

In their recent article, Vitzthum et al. (2005) point out that all specimen types are not equal and often laboratories are forced to analyse the specimens which are made available to them. For example, the choice of plasma over serum may contribute to reducing the potential variability due to differences in clotting time or to products, e.g. the chemokine RANTES which can be released from platelets. Similarly the specimen must be suited to the molecular analysis, e.g. DNA for SNP analysis might best be taken from PMBC obtained from whole blood, while proteomic studies may require plasma or metabolomics studies from urine. More often than not, limited numbers of clinical samples of high integrity are available to laboratories involved in discovery. By way of example, prospective clinical studies, which are designed to provide disease understanding, in the absence of a therapeutic, are costly, slow to recruit and thus are often underpowered.

In developing diagnostic tests, we are faced with another important 'omic'–economic. The cost of developing a diagnostic test to the level of regulatory approval can be in the tens of millions of dollars and represents a considerable risk. In an article on Forbes.com, it was suggested that the extra clinical work to determine if a genetics-based diagnostic test predicts whether a drug will work in a particular patient population could add an extra $100 million to drug development costs. Nonetheless, companies investing in omic approaches to biomarker development, in order to target niche markets, have reaped considerable financial rewards as exemplified by Gleevec (Novartis) and Herceptin (Roche) which generated approximately $1 billion in worldwide sales in 2003. Similarly companies who overlooked potential toxicities have had to pay extensive penalties when successful therapeutics are withdrawn from the market place, e.g. Vioxx (Merck), Baycol (Bayer) and GSK's Lotronex and grepafloxacin. While the economic gains are great and the investment costs are high, the key to any diagnostic effort is substantial evidence of the clinical validity and utility of the test, which then leads to acceptance by caregivers. One negative factor is that while biomarker tests are frequently being developed in laboratory settings, the assay can be expensive to perform. In some cases instrumentation or reagent costs are so high that the evaluation of a large number of assays becomes unfeasible. In addition, factors like the prevalence of the disease and the ability to obtain reimbursement for the test play a part in the economic decisions surrounding the development. In recent articles, Zolg and Langen (2004) and Vitzthum et al. (2005) present a clear picture of how the diagnostic community makes decisions around developability specifically in terms of the phases of diagnostic development.

7 Are the paradigms shifting?

Paradigms for the development of clinical tests focus on establishing validity, a complex characteristic that describes the extent to which a marker reflects a designated event in a biological system. As indicated above, it is the degree of stringency of the validation process which dictates the eventual utility and clinical application of the test. Most of the paradigms suggest three stages of development for the development of clinically valid tests: (1) discovery, (2) verification and validation and (3) establishment of diagnostic viability as a commercial entity. The development of a diagnostic must focus upon tests that allow for definite and reliable diagnosis or those that permit a decision on interventions. In order for a diagnostic to be approved it must meet stringent performance characteristics and have significant sensitivity and specificity. Diagnostic tests have to demonstrate significant positive and negative predictive values which would be expected to depend on disease frequency. Nonetheless, failure of a test to be developed as an approved diagnostic does not limit the value of a product of an 'omic' platform as a biomarker. Perhaps we need to shift the overall paradigm and discuss a spectrum of biomarker qualification rather than validation to the point of regulatory agent approval? In their review, Frank and Hargreaves (2003) point out the important differences between biomarkers and their application to medicine and point out reasons why qualified biomarkers fail as surrogate endpoints. Lesko and Atkinson (2001) present a regulatory view of biomarker evaluation and surrogate endpoints in drug development. They make the point that only a few biomarkers become established enough to serve as surrogate endpoints and suggest that combinations of biomarkers may be useful in characterizing the pharmacological response. From both a regulatory and diagnostic standpoint (Vitzthum *et al.*, 2005) combinational biomarkers represent a challenge with regard to individual analyte validation, the number of analytes which can be combined and data analysis. The question of correlations occurring from chance rather than biological significance arise when multiplexed data are analysed in small sample sets. Lastly there appears to be a need for biomarker development paradigms which are fit for purpose. Barker (2003) describes a 'fit for purpose' paradigm for cancer biomarker validation which includes preclinical, clinical assay development, retrospective longitudinal analysis and prospective screening. Another arena where there is both great promise and exceptional need for data integration and standardization is in the clinical application of microarrays. As clinical researchers begin to use systems like the Roche Amplichip for CYP450, a paradigm establishing best practices for data generation, security and interpretation is required (Jain, 2004; The Tumor Analysis Best Practices Working Group, 2004). Clearly paradigms for biomarker qualification in the realm of metabonomics and molecular imaging will require careful consideration of standards for study design, protocols and instrumentation.

In a recent article, Leigh Anderson suggests that the process of moving the fruit of proteomics platforms towards a clinical diagnostic is thwarted by the lack of an integrated diagnostic development paradigm (Anderson, 2005). Anderson points out that biomarkers are generally identified by academic research and a commercial diagnostics company usually becomes involved late in the process, when there has been some substantiation and acceptance from the academic medicine community. There is ample reason for this lack of integration and it is primarily driven by the differences in cultures between scientists engaged in discovery, those in clinically oriented research and lastly organizations dedicated to the development of diagnostics. The advances in reductionist science have led to remarkable insights into basic mechanisms, predominantly by the use of hypothesis testing in well-defined model systems. In general the culture of discovery-oriented research is one absorbed with *in vitro* and animal model systems. While useful in terms of limiting experimental variability, such model systems are distinctly different from human systems which are characterized by high variability, both at the level of individual patient genetics and the epigenetic factors involved in the evolution of disease. To quote a colleague of mine 'Most hypothesis generating studies are done in relatively small populations, with thousands of potential markers, and of course "predictive" patterns are found ... Consider that you are poling ~30,000 genes in the genome, but a study may only have a few dozen or hundred patients, each of whom has hundreds of phenotypes for various traits. Therefore while you may think you determine an expression pattern for prostate cancer, it could be for brown hair or recent drinking history.'

8 Going rapidly forward

In an earlier review (Bilello, 2005), I suggested that there is a gap in understanding processes such as disease progression (or drug response) which requires insight into dynamic (and temporal) differences in gene regulation, interaction and function. When we look at technological innovation, in translational medicine, we are clearly poised to benefit from a large influx of omic scale information. Will this information lead to a better, more cost-effective approach to develop new therapeutics?

Figure 1 which was put together by some of my colleagues on the Disruptive Technology Innovation team at GSK, describes the journey towards effective treatment of chronic disease or 'wellness' management in terms of a biology axis and engineering axis. On the y-axis, a level of understanding of disease, in all of its various manifestations from biology through drug development to patient issues, ranges from 'simple understanding/simple treatment' (the current situation) to 'complex' total understanding and treatment of complex disease (the desired future state). Similarly, on the x-axis, the application of technologies (e.g. bioengineering, bioinformatics and diagnostics) ranges from the simplistic testing that exists today to more futuristic, dynamic testing done in patients in real time. Quite simply put, we cannot advance without a coming together of disease understanding and technological development.

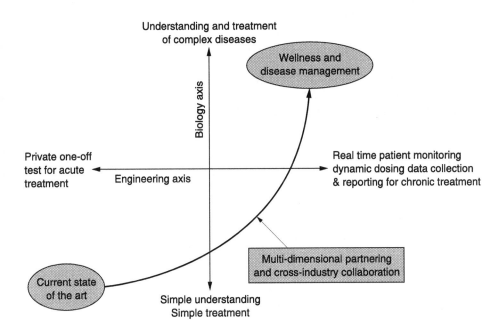

Figure 1.

As indicated in the figure, a multi-dimensional partnership between academics, clinicians, bioengineers, statisticians and computational/systems biologists will be required to generate the platforms, data and tools to provide total health management.

In order to achieve this desirable future state, academics, pharma, diagnostic companies, and governmental agencies will have to find new ways to develop and share 'omic' information and the **economic** costs.

Acknowledgements

I would like to acknowledge the many high level discussions and the continuing support of Dr. Yiwu He, my colleague and co-founder of Precision Human Biolaboratory in Durham, NC. The author would also like to recognize the input of his former colleagues in R&D at Glaxo SmithKline, in particular Dr. Paul Domanico and his Disruptive Technology Initiative team who stimulated the thought processes around biomarkers and personalized medicine.

References

Anderson, L.N. and Anderson, N.G. (2002) The human plasma proteome: history, character, and diagnostic prospects. *Mol Cell Proteomics* **1**: 845–867.

Anderson, N.L. (2005) The roles of multiple proteomic platforms in a pipeline for new diagnostics. *Mol Cell Proteomics* **4**: 1441–1444.

Baker, M. (2005) In biomarkers we trust? *Nature Biotechnol.* **23**: 297–304.

Barker, P.E. (2003) Cancer biomarker validation: standards and process. *Ann NY Acad Sci* **983**: 142–150.

Bilello, J.A. (2005) The agony and ecstasy of 'OMIC' technologies in drug development. *Curr Mol Med* **5**: 39–52.

David, D.C., Hoerndli, F. and Gotz, J. (2005) Functional Genomics meets neurodegenerative disorders Part I: Transcriptomic and proteomic technology. *Prog Neurobiol* **76**: 153–168.

Fan, H. and Hedge, P.S. (2005) The transcriptome in blood: challenges and solutions for robust expression profiling. *Curr Mol Med* **5**: 3–10.

Frank, R. and Hargreaves, R. (2003) Clinical biomarkers in drug discovery and development. *Nat Rev Drug Disc* **2**: 566–580.

Giallourakis, C., Henson, C., Reich, M., Xie, X. and Mootha, V.K. (2005) Disease gene discovery through integrative genomics. *Annu Rev Genomics Hum Genet* **6**: 381–406.

Horig, H., Marincola, E. and Marincola, F.M. (2005) Obstacles and opportunities in translational research. *Nat Med* **11**: 705–708.

Jain, K.K. (2004) Applications of biochips: from diagnostics to personalized medicine. *Curr Opin Drug Discov Devel* **7**: 285–289.

Lesko, L.J. and Atkinson, A.J. (2001) Use of biomarkers and surrogate endpoints in drug development and regulatory decision making: criteria, validation and strategies. *Ann Rev Pharmacol Toxicol* **41**: 347–366.

Lindon, J.C., Holmes, E., Bollard, M.E., Stanley, E.G. and Nicholson, J.K. (2004) Metabonomics technologies and their applications in physiological monitoring, drug safety assessment and disease diagnosis. *Biomarkers* **9**: 1–31.

Man, S.F. and Sin, D.D. (2005) Effects of corticosteroids on systemic inflammation in chronic obstructive pulmonary disease. *Proc Am Thorac Soc* **2**: 78–82.

Morel, N.M., Holland, J.M., van der Greef, J., Marple, E.W., Clish, C., Loscalzo, J. and Naylor, S. (2004) Primer on medical genomics Part XIV: Introduction to systems biology – a new approach to understanding disease and treatment. *Mayo Clin Proc* **79**: 651–658.

Pegram, M. and Slamon, D. (2000) Biological rationale for HER2/neu (c-erbB2) as a target for monoclonal antibody therapy. *Semin Oncol* **27**: 13–19.

Rawlins, M.D. (2004) Cutting the cost of drug development. *Nat Rev Drug Discov* **3**: 360–364.

Rifai, N. and Ridker, P.M. (2001) High-sensitivity C-reactive protein: a novel and promising marker of coronary heart disease. *Clin Chem* **47**: 403–411.

Slamon, D.J., Clark, G.M., Wong, S.G., Levin, W.J., Ullrich, A. and McGuire, W.L. (1987) Human breast cancer: correlation of relapse and survivial with amplification of the HER-2/neu oncogene *Science* **235**: 177–182.

Thadikkaran, L., Siegenthaler, M.A., Crettaz, D., Queloz, P-A., Schneider, P., and Tissot, J.D. (2005) Recent advances in blood-related proteomics. *Proteomics* **5**: 3019–3034.

The Tumor Analysis Best Practices Working Group (2004) Expression profiling – best practices for data generation and interpretation. *Nat Rev Genet* **5**: 229–237.

Tufts Center for Drug Development (2001) Backgrounder: how new drugs move through the development and approval process. Boston, November 2001 (www.tufts.edu/med/csdd/).

US Preventative Services Taskforce (2005) Genetic risk assessment and BRCA mutation testing for breast and ovarian cancer susceptibility: recommendation statement. *Annals Intern Med* **143**: 355–361.

Visser, M., Bouter, L.M., McQuillan, G.M., Wener, M.H. and Harris, T.B. (1999) Elevated C-reactive protein levels in overweight and obese adults. *JAMA* **282**: 2131–2135.

Vitzthum, F., Behrens, F., Anderson, N.L. and Shaw, J.H. (2005) Proteomics: from basic research to diagnostic application. A review of requirements & needs. *J Proteome Res* **4**: 1086–1097.

Watkins, S.M. (2004) Lipomic profiling in drug discovery, development and clinical trial evaluation. *Curr Opin Drug Discov Devel* **7**: 112–117.

Webb, C.P. and Pass, H.I. (2004) Translation research: from accurate diagnosis to appropriate treatment. *J Transl Med* **2**: 35–56.

Yoon, P.W., Chen, B., Faucett, A., Clyne, M., Gwinn, M., Lubin, I.M., Burke, W. and Khoury, M.J. (2001) Public health impact of genetic tests at the end of the 20th century. *Genet Med* **3**: 405–410.

Zolg, J.W. and Langen, H. (2004) How industry is approaching the search for new diagnostic markers. *Mol Cell Proteomics* **3**: 345–354.

7

Drug-target discovery *in silico*: using the web to identify novel molecular targets for drug action

David S. Wishart

1 Introduction

Drug-target discovery is fundamentally a 'wet-bench' experimental process. Most potential drug targets are identified on a laboratory bench top using genetic screens, biochemical tests or cellular assays. These assays may be done in simple cell cultures, but more often are done using a variety of model organisms. While the underlying principles behind *in vitro* drug-target discovery have not changed much in the past 50 years, what has changed profoundly is the throughput or speed by which this process is done. In the past, small numbers of drugs and drug targets were slowly identified through manually intensive and exceedingly tedious laboratory processes. Nowadays large numbers of potential drugs and drug targets are being routinely identified through a variety of high-speed, roboticized technologies including high-throughput DNA sequencing (Carlton, 2003; Kramer and Cohen, 2004), high-throughput microarray or two-dimensional (2D) gel experiments (Butte, 2002; Walgren and Thompson, 2004; Onyango, 2004), rapid-throughput mass spectrometry assays and high-speed roboticized chemical library screens (Jeffrey and Bogyo, 2003; Lindsay, 2003; Comess and Schurdak, 2004).

These same high-throughput technologies that have fundamentally changed the mechanics of 'wet-bench' drug and drug-target discovery are also having a profound change in the way that drug discovery can actually be done. In particular, high-throughput technologies are now allowing – or forcing – drug discovery to be done more on the desk top (i.e. the computer) than on the bench top (i.e. the laboratory). In other words, drug and drug-target discovery are beginning the transition from an *in vitro* process to an *in silico* process.

In silico drug-target discovery is now possible primarily because of the Human Genome Project (Bentley, 2000; Hopkins and Groom, 2002) and related large-scale sequencing efforts. Already more than 2000 viral genomes, 260 bacterial genomes and

Comparative Genomics and Proteomics in Drug Discovery, edited by John Parrington and Kevin Coward. © 2007 Taylor and Francis Group.

more than two dozen eukaryotic genomes have been sequenced and deposited into public databases (http://www.ebi.ac.uk/genomes/). These data are allowing researchers to identify literally thousands of drug targets for both endogenous diseases (central nervous system disorders, diseases of ageing, autoimmune diseases and acquired or in-born metabolic disorders) and infectious diseases (bacterial, viral and parasitic diseases). Many of these targets are being, or can be, rapidly identified *in silico*, using simple sequence comparison and sequence alignment software. Prior to these large-scale genomic sequencing efforts, the total number of endogenous (human) disease genes targeted by all existing drugs was estimated to be less than 400 (Hopkins and Groom, 2002). Now, with all this sequence data in hand, it is estimated that the number of viable endogenous disease drug targets could grow from ~300 to at least 3000 and the number of viable infectious disease drug targets or drug-target classes could grow from ~20 to at least 300 (Hopkins and Groom, 2002).

In silico drug-target identification is not only possible using DNA or protein sequence databases, it is also possible via other kinds of databases and other kinds of software tools. Microarray experiments (Clarke *et al.*, 2004), QTL (quantitative trait loci) mapping (Darvasi, 2005), CGH chip (comparative genomic hybridization chip) experiments (Inazawa *et al.*, 2004), 2D gel experiments (Ryan and Patterson, 2002), ICAT and ITRAC experiments (Patton, 2002), gene knockout and knockdown experiments (Voorhoeve and Agami, 2003; Harris and Foord, 2000), along with NMR-based metabolomic experiments (Griffin, 2003; Fischer, 2005) – all require specialized computer software and databases to make sense of their data and extract the 'critical' genes or biomarker molecules associated with a given disease, a specific mutation or selected biochemical perturbation.

The fact that computer software and computer databases are facilitating or can facilitate so much of today's drug-target discovery is what has partly motivated the preparation of this chapter. Another key catalyst lies in the fact that, unlike wet-bench drug-target discovery, *in silico* drug-target discovery can now be done by just about anyone with a modest computer and a high-speed Internet connection. Indeed a surprising number of high-quality drug discovery and drug-target discovery resources are now freely available over the Internet. In this chapter we will discuss five different classes of web-accessible databases or web-accessible analytical tools and demonstrate how they can be used to identify drug targets *in silico*. These are (1) sequence databases; (2) automated genome annotation tools; (3) text mining tools; (4) integrated drug/sequence databases; and (5) web-based microarray and 2D gel analysis tools. We will briefly survey some of the more established software tools and databases in these categories but will primarily focus on a number of novel tools and databases that have been developed in our laboratory with the explicit intent of facilitating drug-target discovery. For a more complete, and slightly less biased review of other software resources, including a variety of commercial and downloadable products, the following articles are suggested: Fischer (2005), Wishart *et al.* (2005) and Jonsdottir (2005).

2 Defining and identifying drug targets

Before discussing various approaches to *in silico* drug-target discovery, it is important to clarify a few key issues concerning what we regard as drug targets and how they can be identified. While most of us think of drug targets as being genes or proteins, drug targets in fact can be either large molecules (protein, DNA, RNA) or small molecules (metabolites). A survey of drug targets listed through DrugBank (*Table 1*) and the TTD (Chen *et al.*, 2002) indicates that 96% (351/365) of approved drug target

Table 1. Web tools and databases of importance to drug-target discovery.

Sequence databases
GenBank – http://www.ncbi.nlm.nih.gov/BLAST/
Ensembl – http://www.ensembl.org/
UCSC-Genome Browser – http://www.genome.ucsc.edu/cgi-bin/hgGateway
SwissProt – http://us.expasy.org/sprot/
UniProt – http://www.pir.uniprot.org/
GeneCards – http://bioinfo.weizmann.ac.il/cards/index.shtml
Druggable Genome – http://function.gnf.org/druggable/
SymAtlas – http://symatlas.gnf.org/SymAtlas/
OMIM – http://www.ncbi.nlm.nih.gov/entrez/query.fcgi?db=OMIM
HGMD – http://archive.uwcm.ac.uk/uwcm/mg/hgmd0.html
ENTREZ Genomes – http://www.ncbi.nlm.nih.gov/genomes/
EBI Genomes – http://www.ebi.ac.uk/genomes/
BacMap – http://wishart.biology.ualberta.ca/BacMap/

Automated genome annotation tools
PEDANT – http://pedant.gsf.de/
MAGPIE/BLUEJAY – http://magpie.ucalgary.ca/
BASys – http://wishart.biology.ualberta.ca/basys/cgi/submit.pl

Text-mining tools
PubMed – http://www.ncbi.nlm.nih.gov/entrez/query.fcgi
Global Search Engine – http://www.ncbi.nlm.nih.gov/gquery/gquery.fcgi?itool=frompm
MedMiner – http://discover.nci.nih.gov/textmining/main.jsp
MedGene – http://hipseq.med.harvard.edu/MEDGENE/login.jsp
iHOP – http://www.ihop-net.org/UniPub/iHOP/
PolySearch – http://redpoll.pharmacy.ualberta.ca/PolySearch/

Integrated drug/sequence databases
TTD – http://xin.cz3.nus.edu.sg/group/ttd/ttd.asp
PubChem – http://pubchem.ncbi.nlm.nih.gov/
KEGG – http://www.genome.jp/kegg/
PharmGKB – http://www.pharmgkb.org/
DrugBank – http://redpoll.pharmacy.ualberta.ca/drugbank/

Analytical tools
GenePublisher – http://www.cbs.dtu.dk/services/GenePublisher/
GelScape – http://www.gelscape.ualberta.ca:8080/htm/

types are peptide or protein molecules with 93% (757/815) of all non-redundant US Food and Drug Administration (FDA)-approved drugs being targeted to proteins. Another nine drug targets are small molecules (i.e. adenosine, uric acid, digoxin, iduronic acid, asparagine, D-glucosyl-N-acylsphingosine, galactose oligomers, hyaluronic acid and hydroxyapatite), while three classes of DNA (eukaryotic, prokaryotic and viral) and two classes of RNA (bacterial rRNA and retroviral cRNA) serve as nucleic acid drug targets. The ~15 known drugs that target small molecules are almost all protein-based macromolecules (i.e. Raburicase, Pegademase, Hyaluronidase, Digibind, Pancrelipase, Agalsidase beta, Laronidase, Asparaginase, Pegasparagase, Imiglucerase, Aglucerase) with bisphosphonates being perhaps the only small-molecule drugs that target other small molecules. As with protein targets, it is primarily small molecules that target DNA and RNA drug targets (5% or 43/815 of all FDA drugs).

It is also important to distinguish between two general classes of drug targets. Those that are associated with 'endogenous' human diseases and those that are associated with infectious or 'exogenous' diseases. Endogenous diseases are typically chronic human disorders or conditions that arise due to germ-line mutations (genetic diseases), somatic mutations (cancer), the ageing process (atherosclerosis, immune disorders), brain dysfunction, physical or emotional traumas or some other internal factors. Exogenous diseases are typically temporary diseases or conditions that arise from external, non-human agents such as viruses, bacteria, fungi, protozoans, poisons or poisonous animals (snakes, insects). The vast majority of drug targets (97%) and drugs (89%) are associated with endogenous diseases, while only a tiny minority of drug targets (3%) and drugs (11%) are actually associated with exogenous or infectious diseases. This bias is likely due to a number of economic and social factors including the greater profitability of developing drugs to treat chronic (endogenous) conditions, the development of vaccines (to prevent viral infections), improved urban sanitation (which reduced the threat of bacterial diseases) and longer life expectancy.

Interestingly the challenges associated with identifying drug targets and developing drugs to treat endogenous diseases are often far greater than developing drugs to treat infectious or exogenous diseases. This is because most endogenous human diseases have a complex etiology. With the exception of about 400 (Hamosh *et al.*, 2005; Darvasi, 2005) relatively rare, monogenic disorders, the vast majority of endogenous diseases are multi-factorial or polygenic in origin. Added to this is the fact that human physiology is very complex, consisting of hundreds of highly interdependent multi-cellular/multi-organ networks. This makes it particularly difficult to identify disease-causing genes or proteins, regardless of whether one is using *in vivo*, *in vitro* or *in silico* techniques. Illustrated in *Figures 1* and *2* are flow charts outlining how drug-target discovery is typically conducted using *in silico* or *in vivo* techniques. As seen here, there is a close interplay between experimental methods (microarrays, CGH chips, QTL mapping, proteomics) and computational techniques – especially at the initial stages of identifying target lists. As the gene- or protein-target search narrows, one makes greater use of *in silico* tools to select more 'druggable' targets.

Even if a druggable protein or set of proteins is identified, one has the added challenge of creating/finding a drug to make the drug target(s) functional – while not adversely affecting other normal functions. This is a difficult task, especially for a small molecule. As a result, most drugs used to treat endogenous or 'loss-of-function' diseases are simply variants of known human metabolites or small-molecule messengers.

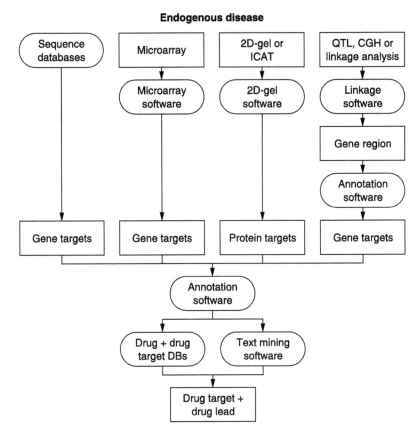

Figure 1. *Flow chart outlining the general process of* in silico *drug-target discovery for endogenous (human) diseases. The rounded boxes represent software tools or databases while the square boxes represent experimental methods.*

While many endogenous diseases arise from a loss of function in a key enzyme or protein, some endogenous diseases or conditions may actually arise from a 'gain in function' or gain in abundance of a particular protein or enzyme or even a small molecule. In these cases, the drug must disrupt or destroy the drug target, but not poison other normally functioning proteins, enzymes or small molecules. This, too, is a very difficult task for a small-molecule drug. Indeed most traditional cancer chemotherapeutics are simply broad-spectrum poisons that attempt to kill the cancer before killing the patient. On the other hand, many of the newer drugs, such as serotonin reuptake inhibitors and selective COX inhibitors, are highly targeted and very effective. Regardless of whether the condition is a 'loss-of-function' or 'gain-in-function' disorder, the process of identifying drug targets in humans is difficult and often requires a combination of wet-bench discovery (2D gels, microarrays and animal knockouts) and desk-top sleuthing (bioinformatics).

In contrast to endogenous diseases, the identification of drug targets and the development of drugs to treat exogenous (especially infectious) diseases is inherently easier and much more compatible with *in silico* approaches (Payne *et al.*, 2004).

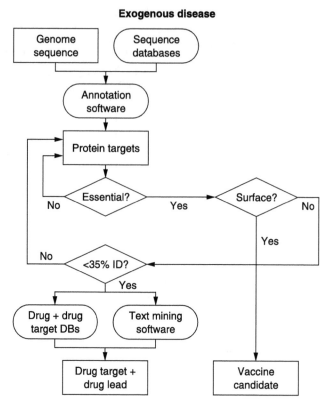

Figure 2. *Flow chart outlining the general process of* in silico *drug-target discovery for exogenous (infectious) diseases. The rounded boxes represent software tools or databases while the square boxes represent experimental methods.*

This process is outlined in *Figure 2*. In many cases (thanks to the availability of high-throughput DNA sequencers) the entire genome of the infectious agent is already known or can be determined in as little as a week. Once the genome is known, the task is then to identify those genes or proteins that are: (1) essential to viability; (2) disease causing; or (3) presented on the surface of the organism. Once these are identified, one must select for those that are sufficiently different (<35% ID) from any human homologues. This prevents any cross-reactivity or potentially adverse drug interactions. Once these non-homologous protein targets are found, one can either search/screen for an inhibitory molecule (for the viability or disease-causing proteins) or develop a vaccine (using parts of the surface proteins). In searching for potential inhibitors, one may search through the literature or existing drug/protein databases to find potential leads based on similarities to previously developed drug candidates or previously known targets.

Overall, the approach to finding drug and vaccine targets for infectious (i.e. exoge-nous) diseases is much simpler and far more compatible with *in silico* methods than for endogenous diseases. As a result, most of the successes in *in silico* target discovery and *in silico* drug development have occurred with viral diseases (i.e. HIV, influenza, hepatitis C – Lahm *et al.*, 2002; Fishman, 1996) and other infectious conditions (Fraser and Rappuoli, 2005; Paine and Flower, 2002). Nevertheless, as techniques and technologies improve, it is likely that *in silico* target discovery will soon demonstrate successes for many endogenous diseases too.

3 Sequence databases

As seen in both *Figures 1* and *2*, the process of *in silico* drug-target discovery often begins with either genomic or proteomic sequence data. Essentially, sequence data represents the 'raw material' for identifying and locating drug targets. The two primary providers of raw (and annotated) sequence data are the National Center for Biotechnology Information (NCBI) in Bethesda (USA) and the European Bioinformatics Institute (EBI) in Hinxton-Head (UK). Both centres have staff levels in excess of 200 individuals and both maintain and operate dozens of general and specialized sequence databases – all of which are freely accessible through their home pages (NCBI – http://www.ncbi.nlm.nih.gov/; EBI – http://www.ebi.ac.uk/). Both the EBI and NCBI are mandated to provide free and up-to-the-minute access to all publicly available sequences (protein and DNA) and their raw sequence repositories are updated and synchronized every 24 hours. As a general rule, the EBI's strength is in providing protein-rich resources while the NCBI's strength is in providing DNA-rich resources (Brooksbank *et al.*, 2005; Wheeler *et al.*, 2005).

Within the NCBI's and EBI's database collections are a number of 'whole genome' resources. These are organism-specific databases containing modestly annotated sequence (protein and DNA) data. In addition to these NCBI and EBI resources, there are a growing number of richly annotated organism-specific or kingdom-specific databases (all derived from the raw data at NCBI or EBI) that can serve as equally good discovery vehicles (Frazer *et al.*, 2003; Stothard *et al.*, 2005; Riley *et al.*, 2005). Drug targets may be identified by simply 'trawling' through these databases, looking for interesting proteins with desired names or appropriate functions. This can be done for the entire genome (for viruses or bacteria), selected genes or proteins (from microarray or 2D gel experiments) or for a defined chromosomal region that has been identified through experimental QTL or linkage mapping (for endogenous human diseases). Alternatively drug targets, especially for endogenous diseases, may be identified simply by comparing two sequences, one being the suspected disease gene/protein and the other being the wild-type. If sequence differences are found, then it is often a strong indication that a potential drug target has been identified. When identifying bacterial or viral targets, one can also compare the sequences of pathogenic versus non-pathogenic forms, allowing the identification of so-called pathogenicity islands (Buysse, 2001; Chan *et al.*, 2002). These islands or clusters of genes typically code for toxins or infectivity proteins that are unique to the given pathogen. These pathogenically unique proteins can serve as particularly good and highly selective drug targets. There are obviously many other ways that these sequence databases can be used to identify drug targets, all of which depend on the type of knowledge or experimental data one has and the type of disease (endogenous or exogenous) one is targeting.

4 Sequence databases for endogenous diseases

When identifying or homing in on endogenous disease targets, one is most often interested in human gene or protein sequences. However, with recent developments in comparative genomics and comparative proteomics using model organisms, there is a growing realization that sequence information from other organisms can also be put to very good use (Reiter and Bier, 2002; Zon and Peterson, 2005). There are at least five sequence databases, containing both human and model metazoan sequence data, that are particularly useful in identifying drug targets for endogenous diseases. These include: Ensembl (Hammond and Birney, 2004; Brooksbank *et al.*, 2005), the UCSC Genome Browser (Karolchik *et al.*, 2003), EntrezGene (Wheeler *et al.*, 2004), GeneCards (Rebhan *et al.*, 1998) and SwissProt (O'Donovan *et al.*, 2002). Website addresses for all of these databases are given in *Table 1*. Ensembl is both a queryable database and a web-accessible genome viewer that contains the sequence data and automatic annotation and schematic image maps of numerous metazoan genomes. Ensembl also contains information about gene organization or gene order (synteny), chromosome structure and sequence relationships between different genes in different genomes (orthology). The Ensembl database is continuously updated with the human genome being updated every two months. Just like Ensembl, the NCBI genome resources (especially EntrezGene) also support web-accessible queries for a number of other metazoan genomes. Both Ensembl and EntrezGene depend on reference sequence (RefSeq) data supplied by the NCBI. RefSeq is a comprehensive, publicly available, non-redundant set of sequences, including genomic DNA, transcript (RNA) and protein products, for all major research organisms (Wheeler *et al.*, 2005).

The GeneCards database (Rebhan *et al.*, 1998) is perhaps the most comprehensive biomedical/gene sequence resource for the human genome. It is primarily a web-queryable database of human genes, their products and their involvement in diseases. The GeneCards database is regenerated and updated every 2–3 months by continuously data mining a large number of public databases – including SwissProt, Ensembl and RefSeq. A standard GeneCard includes information on the gene's official (HUGO) name, synonyms, gene IDs, UniGene cluster, cytogenetic locus, known SNPs, gene coordinates, the name of the gene products, their cellular functions, gene expression graphs, similarities with other proteins, involvement in diseases, orthologues to mouse genes, relevant references and a list of disorders and mutations in which the gene is involved. These data are obviously quite relevant to any researcher working in the area of drug or drug-target discovery.

SwissProt (O'Donovan *et al.*, 2002) is a manually curated, protein-only sequence database that provides a very high level of annotation about protein sequences, names, functions, enzymatic reactions, drug interactions (occasionally), properties and relationships. SwissProt is updated every 4–6 months and has been recently integrated into UniProt (the Universal Protein Resource) – a much larger database containing the TrEMBL protein sequence database (Bairoch *et al.*, 2005). The most recent release of SwissProt contains 195,489 sequence entries, comprising 70,852,380 amino acids abstracted from 134,391 references. SwissProt (release 48.2) contains sequence and sequence annotation information from more than 9495 different species including 12,946 human sequences and 10,034 mouse sequences. A typical SwissProt entry contains information on the protein name, synonyms, gene name, taxonomy, references, probable function, subunit structure, cellular location, reactions, catalytic activity, polymorphisms, substrates/products, EC numbers, tissue specificity, induction, similarity, sequence features, membrane spanning regions (if any), signal peptides (if any), sequence conflicts, sequence variants, 2D PAGE location(s), links to structure, links to sequence motifs, links to protein interactions or interacting partners, sequence, sequence length and molecular weight. The SwissProt database is regarded by many as the gold standard for protein annotation and protein sequence information.

In addition to these five well-known sequence databases, there are at least four other smaller (some of which are essentially non-sequence) databases that are particularly useful in endogenous disease target identification. These are the Druggable Genome database (Orth *et al.*, 2004), SymAtlas (Orth *et al.*, 2004), OMIM (Hamosh *et al.*, 2005) and HGMD (Stenson *et al.*, 2003). The Druggable Genome database contains links, BLAST search tools and raw sequence data derived from studies by several groups in the Genomics Institute of the Novartis Research Foundation (GNF) pertaining to the 3000 or so human genes that appear to represent good or potentially 'druggable' drug targets using the rules of Hopkins and Groom (2002). This database also contains links to several other GNF software packages or databases including GeneAtlas and SymAtlas. SymAtlas is a GeneCards-like database that supports queries over several different genomes (human, mouse, rat and other metazoans), but its primary focus is on human data. While generally providing less comprehensive annotation data than GeneCards, SymAtlas is quite notable in that it provides Affymetrix (U133A) gene chip expression data for many gene entries covering nearly 80 different human tissues or cell types. Because these data were collected by a single laboratory on a uniformly prepared set of tissues, this data resource offers a unique opportunity for drug researchers to look at tissue-specific or organ-specific gene expression levels. This information is particularly important in light of the tremendous variation in abundance and expression for many genes and proteins amongst different organ systems. Obviously one does not want to spend an enormous amount of effort identifying a likely drug target (say in cell culture or through *in vitro* assays), only to find it is not expressed in the organ that needs to be targeted. SymAtlas could certainly help prevent mistakes like these.

Drug target identification is certainly much easier if the disease-causing gene(s) are already formally identified. The On-line Mendellian Inheritance of Man (OMIM) is one such resource that provides this kind of information. OMIM is a superbly researched encyclopaedic resource containing genetic (cloning, gene function, gene structure, mapping), phenotypic, historical and clinical data on more than 5500 genetic and/or metabolic disorders (Hamosh *et al.*, 2005). Of these, nearly 400 diseases have specific gene sequences associated with them and another 1600 disorders have a solid understanding of their molecular origins. Many of the key, disease-causing mutations in OMIM are described in detail and are linked to either NCBI sequence data or PubMed references. OMIM is searchable by title, text, clinical data, allelic variants, chromosome number and citations. It is also fully downloadable.

While not as encyclopaedic as OMIM, the human gene mutation database (HGMD) is another pharmaceutically useful database that links human diseases to genes or at least mutations in genes. The HMGD is primarily a repository of sequence mutations found in well-studied human genes. The HGMD contains 47,889 mutations compiled for 1885 genes with 1748 reference cDNA sequences. The database can be searched by disease, gene name or gene symbol. A more complete description of HGMD is given by Stenson *et al.* (2003).

5 Sequence databases for exogenous diseases

When identifying exogenous disease targets, one is primarily interested in processing bacterial or viral sequences. There are at least five databases that are particularly useful in this regard: NCBI's Microbial and Viral Genome resources or 'Entrez Genomes' (Wheeler *et al.*, 2005), EBI's Viral and Microbial Genome resources (Brooksbank *et al.*, 2005), TIGR's Comprehensive Microbial Resource (Peterson *et al.*, 2001), BacMap (Stothard *et al.*, 2005) and the Viral Genome Database (Hiscock and Upton, 2000). Combined, these resources contain more than 260 different bacterial genomes (of which more than 40% belong to known pathogenic forms) and more than 2000 viral genomes (covering most known viral pathogens). Currently, one new bacterial genome is being added to these databases every week and one new viral genome is being added every day. Genome sequence data are also available for a number of pathogenic fungi (candida, aspergillus), and now several protozoan parasites (guillardia, leishmania, *Plasmodium falciparum*). Almost all of these microbial/viral genome databases are searchable or browseable via species name, general text or sequence queries. For the NCBI's Entrez Genomes, complete sets of microbial or viral genomes can be accessed hierarchically starting from either an alphabetical listing or a phylogenetic tree. From there, it is possible to follow the hierarchy to a graphical overview for the genome, down to the level of a single chromosome and, finally, down to the level of a single gene. At each level are one or more views, pre-computed summaries and links to analyses. At the level of a genome or a chromosome, a viewer displays the location of each coding region, length of the product, GenBank ID for the protein sequence and protein name. At the level of a single gene, links are provided to pre-computed sequence neighbours for the implied protein with links to the COGs (clusters of orthologous genes) database if possible. A summary of COG functional groups is presented in both tabular and graphical formats at the genome level. For complete microbial genomes, pre-computed BLAST neighbours for protein sequences and links to their 3D structures are also provided. Pairwise sequence alignments are presented graphically and linked to the Cn3D macromolecular viewer. A new tool called GenePlot allows genome-wide comparisons of protein homologies to be visualized so that genomic inversions, deletions and insertions between bacterial strains and closely related species can be identified and highlighted.

Given the tendency of most bacterial databases to be somewhat textual and relatively geno-centric, this can potentially limit a more detailed 'visual' or functional exploration. The need to be able to visualize bacterial genomes and to be able to link genomic data to more detailed proteomic data is particularly important in understanding bacterial pathogenicity and in identifying potential drug targets (Payne *et al.*, 2004; Paine and Flower, 2002). Given the utility of visual tools in drug-target discovery, my laboratory decided to develop a more comprehensive and visually oriented database called BacMap (Stothard *et al.*, 2005). BacMap is essentially an electronic atlas of bacterial genomes modelled after the Ensembl browser. It contains hundreds of circular, multi-coloured, clickable maps, all linked to thousands of megabytes of detailed annotation. When the user enters the BacMap site *(Table 1)*, the entry page displays a scrollable list of all publicly released prokaryotic genomes (>250). This list is presented alphabetically, according to genus, species and strain. Each bacterial name is hyperlinked to a 'species card' which provides detailed information about the organism in tabular format, including its taxonomy, Gram staining properties, and number of chromosomes. A brief description of the species, discussing its physiology, general characteristics, ecological niche and relevance to human or animal disease, is also given. Below each genome entry is a list of the genome's constituent chromosomes, sorted by length. Five buttons are provided for each chromosome: 'Map', 'Text Search', 'BLAST', 'Stats' and 'Download'. The 'Map' button displays a graphical map of the chromosome in a new window. The 'Text Search' and 'BLAST' buttons are linked to the text search and BLAST search interfaces for the chromosome. The 'Stats' button displays several graphs concerning the chromosome's genomic and proteomic characteristics. Finally, the 'Download' button opens the data download page for the chromosome.

Clicking on the 'Map' button generates a full-screen circular image of the entire bacterial chromosome. On the lower edge of the image is a brief synopsis of the chromosome, including the GenBank accession number of the source sequence. The full view map consists of two concentric rings of forward and reverse strand genes (protein and RNA), with tick marks indicating chromosomal position. Some maps may contain additional feature rings (COG functional classifications for example) depending on which annotations are currently available for the chromosome. Clicking on any of the tick marks in the sequence ruler expands the map by a pre-defined step, and centres the view on the base closest to the tick mark that was clicked. The map view can also be manipulated using the control panel located at the bottom of the map. Hyperlinked gene labels are visible from the first zoom level onwards. Pointing to a gene label displays the start and stop positions of the gene, as well as its known or predicted function. Clicking on the gene label replaces the map view with the corresponding 'BacMap card'. The chromosome maps can be explored manually, or with the assistance of two search tools integrated with the BacMap database. One tool is a Boolean text search, the other is a sequence search – BLAST (Altschul et al., 1997). As with the text search, the BLAST results are shown graphically and textually, with hyperlinks provided for accessing the chromosome maps and gene cards. *Figure 3* provides a montage of BacMap images to demonstrate the image quality, utility and general operation of the BacMap database.

Relative to other microbial sequence databases, BacMap is quite richly annotated. These annotations, presented in a simple tabular format, are accessible by clicking on the gene labels displayed on individual chromosome maps, or by using the text and BLAST searches. The information contained in the BacMap cards is built using data found in a variety of public databases, such as UniProt (Bairoch et al., 2005) and PDB (Deshpande et al., 2005), as well as numerous in-house prediction programs. Each BacMap card typically contains 40+ fields of annotation, including information on a variety of sequence statistics, potential orthologues and paralogues, predicted function, predicted secondary structure, predicted subcellular location and predicted 'essentiality' (critical for the organism's survival). The information on protein essentiality is particularly important if one is trying to identify potential drug targets in microbes. Likewise the ability to compare or visualize multiple genomes and to look for clusters of genes (operons, pathogenicity islands) opens the door to more rapid or targeted evaluation of potential microbial gene targets.

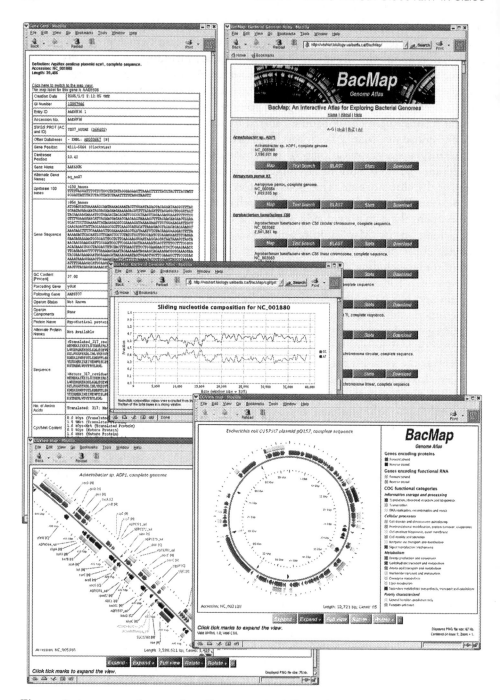

Figure 3. *A montage of screen shots illustrating the different functions and views available for genomes rendered by BacMap.*

6 Automated genome annotation tools

Sometimes in drug discovery, the genome of interest is not known and it is necessary
to sequence a new strain or variant of the virus or bacterium of interest. In these cases
one only has the genomic sequence in hand, and little else. In other situations, the
genome is already sequenced but the annotations are inadequate, outdated or incomplete. This is true even for the human genome (Southan, 2004). In other words, one
cannot always depend on the information in existing sequence databases to find
a drug target of interest. To deal with these database limitations a number of biomedical researchers are turning towards automated genome annotation tools. Genome
annotation tools allow scientists to either re-process existing sequence data or process
new sequence data. This allows them to generate the up-to-date annotations they
need to do their research. Over the years a number of automated genome annotation
tools have been created, including GeneQuiz (Andrade *et al.*, 1999), PEDANT (Riley
et al., 2005), Genotator (Harris, 1997), MAGPIE/BLUEJAY (Gaasterland and
Sensen, 1996; Turinsky *et al.*, 2005), GenDB (Meyer *et al.*, 2003) and the TIGR CMR
(Peterson *et al.*, 2001). Unfortunately, because of their complexity most of these
genome annotation systems are proprietary and typically run on internally maintained systems. This obviously makes it difficult for external users to perform routine
annotation updates or to generate *de novo* annotations on their own.

However, this problem is slowly being overcome with the development of several
publicly available genome or proteome annotation servers (Szafron *et al.*, 2004). One
such example is BASys (Van Domselaar *et al.*, 2005) which was developed in our laboratory over the past three years. BASys (Bacterial Annotation System) is a publicly
available web server designed to support automated, in-depth annotation of bacterial
genomic (chromosomal and plasmid) sequences. BASys accepts either raw DNA
sequence data (whereupon it identifies the genes using GLIMMER [Delcher *et al.*,
1999]) or pre-determined gene sequence lists. If an ordered list of gene or cDNA
sequences is already available, BASys can be 'fooled' into processing and annotating
viral and eukaryotic sequences as well. After receiving the appropriate sequence input
BASys uses more than 30 different programs to determine nearly 60 annotation
subfields for each gene, including gene/protein name, GO function, COG function,
possible paralogues and orthologues, molecular weight, isoelectric point, operon
structure, subcellular localization, signal peptides, transmembrane regions, secondary
structure, 3D structure, reactions and pathways. A detailed summary of the annotation process for each field along with its relevance and the program or database ver-
sions used is available on the BASys web site under the annotations link (*Table 1*).
In general, the depth and detail of a BASys annotation generally matches or exceeds
that found in a standard SwissProt entry. BASys also generates colourful, clickable and
fully zoomable maps of each query chromosome to permit rapid navigation and
detailed visual analysis of all resulting gene annotations. The textual annotations and
images that are provided by BASys can be generated in approximately 24 hours for an
average bacterial chromosome (5 megabases). BASys annotations may be viewed and
downloaded anonymously or through a password protected access system. A montage
illustrating the various types of output produced by BASys is provided in *Figure 4*.

BASys' unique integration of operon, reaction and pathway information into the
annotation process is particularly useful when identifying infectious disease drug targets.

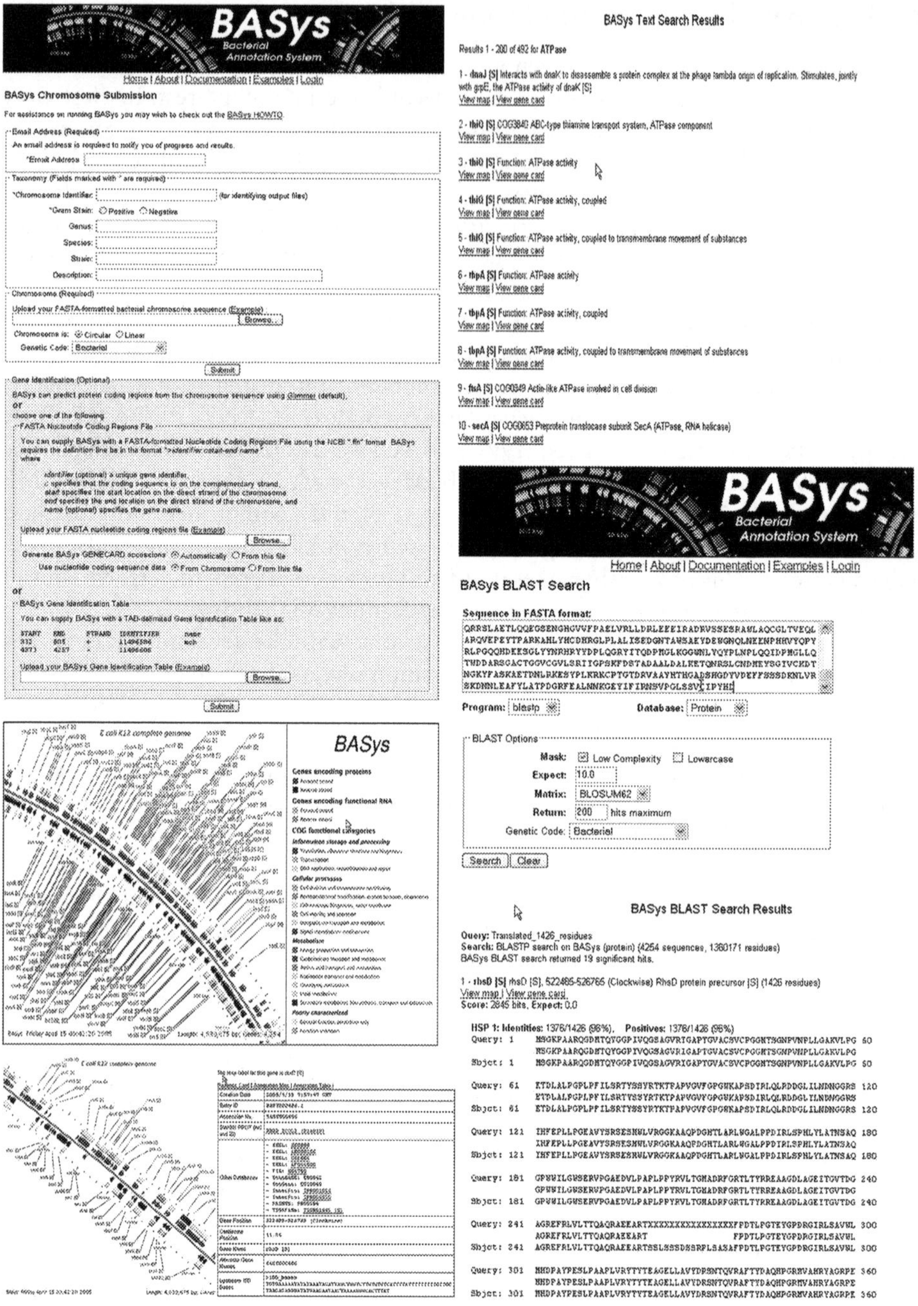

Figure 4. A montage of screen shots illustrating the annotations generated by BASys.

Sometimes the ideal drug target is not the disease-causing protein itself, but rather the upstream or downstream enzymes that are involved in the same signalling or meta-bolic pathway. Other times the best drug target may prove to be the small-molecule products of the enzyme of interest rather than the enzyme itself. Likewise, hints about the 'druggability' of a proposed target can often be gleaned from the type of small-molecule substrates or cofactors that the protein of interest binds and/or modifies.

While BASys is generally limited to bacterial annotation, it can be used to annotate other kinds of proteomes (viral, eukaryotic). However, these annotations will have a certain bacterial 'flavour'. Certainly the development of similar kinds of automated, web-based annotation systems for eukaryotic organisms would be a welcome addition – especially for those interested in endogenous disease target identification.

7 Text-mining tools

In some cases sequence data and sequence annotation is simply not enough to permit *in silico* drug-target discovery. Indeed, there is always a lag between what appears in public sequence databases and sequence annotation systems and what appears in the scientific literature. New protein interactions might be found, new functions may be discovered, new pathways could be identified, new relationships between a given gene and a given disease may be uncovered, novel drug leads could be reported or newer and better algorithms may be developed – all of these advances are typically reported in the literature six months to two years before they make it into any sequence data-base or sequence annotations system. Furthermore, many new and pharmaceutically important pieces of data are simply not accommodated in the rigidly defined data fields of most sequence databases or annotation systems. All this is really saying is that to stay at the cutting edge of drug discovery, one must stay current with the lit-erature.

However, staying current with the literature can be quite a challenge. It has been estimated that in order for a scientist to stay current for a single high-priority disease (say breast cancer), they would have to scan 130 different journals and read 27 papers each week (Baasiri *et al.*, 1999). With each article being approximately eight journal pages each or about 3000 words, this represents about 15,000 words or 50 pages a day. Given that most journal articles are not exactly 'light' reading, this task of staying cur-rent with the literature could easily occupy 75% of a scientist's working day. Obviously most of us are not so diligent. Fortunately, there are a number solutions – both existing and on the horizon – that will make staying current with the literature much easier. Most of these advances have to do with a computational field called text mining.

Biological text mining is an active area of bioinformatics research with a growing number of publications and a growing number of computational approaches available to users (Chaussabel, 2004; Cohen and Hersh, 2005). Almost all biomedical text mining systems use the wealth of biological and chemical abstract information now available on-line, through PubMed. PubMed, which is maintained by the National Library of Medicine, contains more than 12.8 million abstracts from 4400 biomedical and biochemical journals dating to as far back as the 1970s (Wheeler *et al.*, 2005). As a web-based text-mining tool, PubMed was perhaps the first to appear and probably remains as the most popular of all text-mining systems. Indeed, PubMed has essentially inspired and enabled much of the current work in biomedical text mining. However, PubMed is somewhat limited in its search capabilities allowing only Boolean text queries on MESH (Medical Subject Headings) words. It does not support such advanced features as phrase selection or sentence context selection. More recently PubMed has become more fully integrated with NCBI's Entrez Cross-Database searchsystem (Wheeler *et al.*, 2005) so that users can see more than just journal abstracts and titles to their text queries. The New Entrez Global Search Engine will also return links to relevant or matching DNA and protein sequence files in GenBank, chemical structures in PubChem, gene expression data from GEO and 3D structures from the PDB and Entrez Structure. These new enhancements certainly could make Entrez a very powerful tool for *in silico* drug and drug-target discovery.

In addition to PubMed and Entrez's Global Search Engine, several other very useful and far more advanced on-line web tools are available. These include MedMiner (Tanabe *et al.*, 1999), iHOP (Hoffmann and Valencia, 2005), MedGene (LaBaer, 2003) and PolySearch (*Table 1*). While still dependent on the information contained within the PubMed database (and the PubMed applications programming interface (API) for submitting multiple queries), these web servers also support more sophisticated text and phrase searching, phrase selection and relevance filtering.

MedMiner (Tanabe *et al.*, 1999) is a web-based text search tool that filters, extracts and organizes relevant sentences in the literature based on a gene, gene–gene or gene–drug query. MedMiner combines the GeneCards and PubMed search engines with user input and automated server-side scripts in an integrated text filtering system. MedMiner first requires a user to specify the genes of interest either by entering specific gene names or by entering a general concept, disease or disorder (e.g. breast cancer) that can be used to find genes. The gene names or concepts are then sent as a query to the GeneCards database. For gene/drug queries, MedMiner uses drug synonyms obtained from the NCI's Drug Information System database. After retrieving the relevant information from the GeneCards database, the gene, drug and/or disorder names are then formulated by MedMiner into a PubMed query. In doing so, MedMiner creates a Boolean PubMed search with user-determined combinations of synonyms and retrieves the citations of the matching articles. The results are then filtered using different 'relevance' metrics or 'relevance' filters. For instance, a sentence is considered relevant if it contains at least one gene synonym and at least one keyword. A citation is considered relevant if its title or abstract contains at least one relevant sentence. As an example, an abstract might be considered relevant if it contains a sentence with both the name of the gene and the word 'inhibits'. MedMinder also uses word frequencies to determine relevance. A frequency-based filter might specify that an abstract is relevant if it contains words like 'gene', 'inhibit' or 'inhibition' significantly more frequently than does an average document. After performing the filtering steps, MedMiner generates a results page that shows the Boolean PubMed query, summary statistics on the number of abstracts and sentences found, links to possible false positives and a set of hyperlinked tables. Each citation entry is annotated with the sentence that explains its relevance with the keywords highlighted, and a link to the unfiltered abstract is provided. Note that the close linkage between drugs, human diseases and human genes makes MedMiner a particular useful text-mining tool for finding drug targets, as well as drug leads, especially for endogenous diseases.

MedGene (LaBaer, 2003) is a database maintained at Harvard and built from an automated literature-mining tool that comprehensively summarizes and estimates the relative strengths of all human gene–disease relationships in Medline. MedGene uses statistical methods on gene–disease co-citations from titles, abstracts and MESH terms in order to rank the relative strengths of gene–disease association. MedGene allows users to search or review data on six different categories including: (1) a list of human genes associated with a particular human disease in ranking order; (2) a list of human genes associated with multiple human diseases in ranking order; (3) a list of human diseases associated with a particular human gene in ranking order; (4) a list of human genes associated with a particular human gene in ranking order; (5) a sorted gene list from other disease-related high-throughput (i.e. microarray) experiments; and (6) a sorted gene list from other gene-related high-throughput (i.e. microarray) experiments. MedGene tries to encompass all reported gene–disease links, including genetic, biochemical, pharmacological, epidemiological and physiological connections. It also assigns a mathematical score summarizing the strength of the association between the disease and the gene, which allows semiquantitative analysis of the results. While not explicitly linked to drug data, MedGene does offer a fast and powerful approach to identifying potential drug targets, especially for most endogenous diseases.

MedMiner and MedGene are both excellent text-mining tools. However, they are still somewhat limited in their capabilities as they were developed with an assumption of users having fairly substantial domain-expertise and the requirement that user's provide constant input. In an effort to create a more automatic, user-naive text-mining tool – especially for applications in drug-target discovery – we developed PolySearch. PolySearch (Polymorphism Search – *Table 1*) is a web-accessible resource that supports PubMed literature searches on drugs, disease genes as well as gene name searches, SNP searches, mutation searches and PCR-primer searches. It differs from MedMiner and MedGene in the number of search possibilities, its general search strategies, its support for comparative genomics research (SNPs) and its search speed (it is more automatic, and much faster). PolySearch also allows users to design or generate primers for polymorphism or mutational analysis from the gene/SNP/mutation data that it has mined. Users may search through a variety of SNP or mutation databases (HGVbase, dbSNP, HGMD) and text resources (PubMed, lists of disease names/synonyms, gene names/synonyms, drug names/synonyms). A key strength to PolySearch is the extensive and up-to-date lists it maintains of disease synonyms (4000+), protein/gene synonyms (40,000+), and drug synonyms (14,000+). Each type of search may be performed independently (e.g. find all genes associated with tamoxifen; find all polymorphisms and mutations associated with genes adrl, drcl and trxA; find PCR primers for gene sequences adrl, drcl and trxA) or in a combined fashion (find all SNPs for all genes associated with breast cancer and design all necessary primers for subsequent SNP analysis). PolySearch uses a variety of techniques including text mining, web-based screen scraping, and primer design to generate its results, which can be sent as an HTML hyperlinked table or presented in a web-accessible relational (mySQL) database format. PolySearch is specifically designed to allow pharmaceutical researchers to explore the relationships between drugs, polymorphisms and/or diseases and physiological responses. A screenshot of several PolySearch web pages is shown in *Figure 5*.

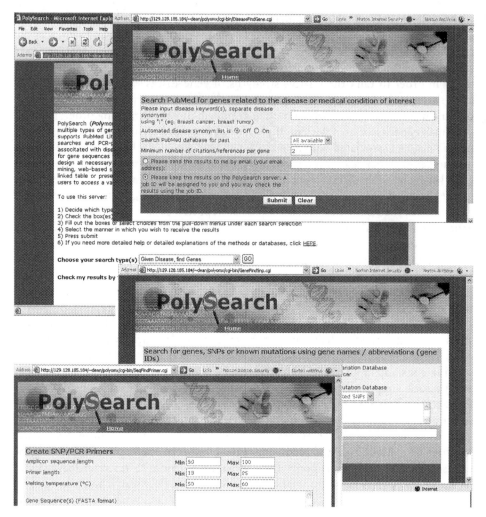

Figure 5. *A montage of screen shots illustrating the different functions and capabilities found in PolySearch.*

8 Integrated drug/sequence databases

Most sequence databases were developed without the intention of using this data to facilitate drug-target discovery. As a result most sequence data are not linked in any meaningful way to existing drug or disease information. This 'information disconnect' is one of the reasons why bioinformatics has been so slow to help in the drug discovery and drug-target discovery process. By constructing comprehensive, meaningful links between drugs, diseases and sequences, it should be possible to learn from past successes and even past failures in terms of what proteins make for good drug targets (soluble vs. membrane-bound, structural proteins vs. enzymes, strong binders vs. weak binders), what types of pathways (metabolic, signalling, nuclear, cytoplasmic) make for good therapeutic intervention strategies, what characteristics in small molecules make for good drug leads (Lipinski's rule of five), what classes of drug targets are under or over-represented in existing formularies, and so on. Generally this information was available only to pharmaceutical companies or to research facilities that were wealthy enough to afford certain proprietary databases (Hopkins and Groom, 2002). This 'cost barrier' largely prevented academic research groups from exploring newly sequenced genomes with the intent of finding new drug targets and drug leads.

Attempts are now being made to remedy this situation. For instance, the NCBI has now integrated OMIM (disease information), GenBank (sequence information) and PubChem (chemical or drug information) into its freely available Entrez Global Search Engine (Wheeler *et al.*, 2005). Other efforts are also underway including GNF's Druggable Genome database (Orth *et al.*, 2004) and the Therapeutic Target Database or TTD (Chen *et al.*, 2002). The TTD is a freely accessible web-based resource that contains linked lists of names for more than 1100 small-molecule drugs and drug targets (e.g. proteins). It contains information about known protein and nucleic acid targets together with the associated disease conditions, pathway information and the corresponding drugs/ligands directed to each drug target. Hyperlinks to other databases facilitate access to information regarding the function, sequence, 3D structure, nomenclature, drug/ligand binding properties and related literature about each protein/DNA target.

In addition to the TTD, a number of comprehensive small-molecule databases have also emerged including KEGG (Kanehisa *et al.*, 2004), ChEBI (Brooksbank *et al.*, 2005) and PubChem (Wheeler *et al.*, 2005). Each contains tens of thousands of chemical entries – including hundreds of small-molecule drugs. All three databases provide names, synonyms, images, structure files and hyperlinks to other databases. Furthermore, both KEGG and PubChem support structure similarity searches. Unfortunately, these databases were not specifically designed to be drug databases, and so they do not provide specific pharmaceutical information or links to specific drug targets (e.g. sequences). Furthermore, because these databases were designed to be synoptic (containing fewer than 15 fields per compound entry) they do not provide a comprehensive *molecular* summary of any given drug or its corresponding protein target. More specialized drug databases such as PharmGKB (Hewett *et al.*, 2002) or on-line pharmaceutical encyclopaedias such as RxList (Hatfield *et al.*, 1999) tend to offer much more detailed clinical information about many drugs (their pharmacology, metabolism and indications) but they were not designed to contain structural, chemical or physicochemical information. Instead their data content is targeted more towards pharmacists, physicians or consumers – not drug-target discovery specialists.

In an effort to create a single, fully searchable *in silico* drug resource that links sequence, structure and mechanistic data about drug molecules with sequence, structure and mechanistic data about their drug targets, we have developed a new database called DrugBank. Fundamentally DrugBank is a dual-purpose bioinformatics–cheminformatics database with a strong focus on quantitative, analytic or molecular-scale information about both drugs and drug targets. In many respects it combines the data-rich molecular biology content normally found in curated sequence databases such as SwissProt and UniProt (Bairoch *et al.*, 2005) with the equally rich data found in medicinal chemistry textbooks and chemical reference handbooks. By bringing these two disparate types of information together into one unified, freely available resource it should allow a much larger community to conduct *in silico* drug and drug-target discovery.

DrugBank currently contains more than 4100 drug entries, corresponding to >14,000 different trade names and synonyms. To facilitate more targeted research and exploration, DrugBank is divided into four major categories: (1) FDA-approved small-molecule drugs (>700 entries), (2) FDA-approved biotech (protein/peptide) drugs (>100 entries), (3) nutraceuticals or micronutrients such as vitamins and metabolites (>60 entries) and (4) experimental drugs, including unapproved drugs, de-listed drugs, illicit drugs, enzyme inhibitors and potential toxins (3200 entries). These individual 'Drug Types' are also bundled into two larger categories including all FDA drugs (Approved Drugs) and All Compounds (experimental + FDA + nutraceuticals). DrugBank's coverage for non-trivial FDA-approved drugs is approximately 80% complete.

DrugBank is fully searchable with many built-in tools and features for viewing, sorting and extracting drug or drug target data. As with any web-enabled database, DrugBank supports standard text queries (through the text search box located on the home page). It also offers general database browsing using the 'Browse' and 'PharmaBrowse' buttons located at the top of each DrugBank page. To facilitate general browsing, DrugBank is divided into synoptic summary tables which, in turn, are linked to more detailed 'DrugCards' – in analogy to the very successful GeneCards concept (Rebhan *et al.*, 1998). All of DrugBank's summary tables can be rapidly browsed, sorted or reformatted in a manner similar to the way PubMed abstracts may be viewed. Clicking on the DrugCard button found in the leftmost column of any given DrugBank summary table opens a webpage describing the drug of interest in much greater detail. Each DrugCard entry contains more than 80 data fields with half of the information being devoted to drug/chemical data and the other half devoted to drug target or protein data. In addition to providing comprehensive numeric, sequence and textual data, each DrugCard also contains hyperlinks to other databases, abstracts, digital images and interactive applets for viewing molecular structures (*Figure 6*).

A key feature that distinguishes DrugBank from other on-line drug resources is its extensive support for higher level database searching and selecting functions. In addition to the data viewing and sorting features already described, DrugBank also offers a local BLAST (Altschul *et al.*, 1997) search that supports both single and multiple sequence queries, a Boolean text search (using GLIMPSE – Manber and Bigot, 1997), a chemical structure search utility and a relational data extraction tool (Sandararaj *et al.*, 2004). These can all be accessed via the database navigation bar located at the top of every DrugBank page. The BLAST search (SeqSearch) is particularly useful as it can potentially allow users to quickly and simply identify drug-target leads from newly sequenced pathogens (i.e. exogenous diseases). Specifically, a new sequence, a group of sequences or even an entire proteome can be searched against DrugBank's database of known drug target sequences by pasting the FASTA formatted sequence (or sequences) into the SeqSearch query box and pressing the 'submit' button. A significant hit reveals, through the associated DrugCard hyperlink, the name(s) or chemical structure(s) of potential drug leads that may act on that query protein (or proteome).

DrugBank's structure similarity search tool (ChemQuery) can be used in a similar manner to its sequence search tools. Users may sketch or paste a SMILES string (Weininger, 1998) of a possible lead compound into the ChemQuery window. Submitting the query launches a structure similarity search tool that looks for common substructures from the query compound that match DrugBank's database of known drug or drug-like compounds. High-scoring hits are presented in a tabular format with hyperlinks to the corresponding DrugCards (which in turn links to the protein target). The ChemQuery tool allows users to quickly determine whether their compound of interest acts on the desired protein target. This kind of chemical structure search may

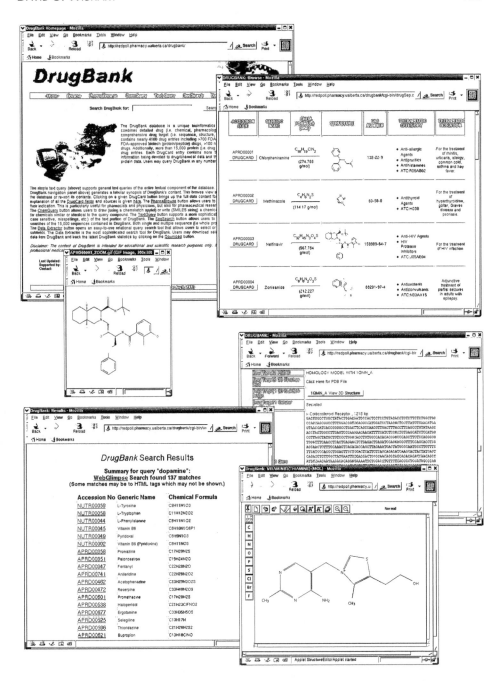

Figure 6. A montage of screen shots illustrating the different functions, fields and views available in DrugBank.

also reveal whether the compound of interest may unexpectedly interact with unintended protein targets. In addition to these structure similarity searches, the ChemQuery utility also supports compound searches on the basis of chemical formula and molecular weight ranges.

Overall, DrugBank is a comprehensive, web-accessible database that brings together quantitative chemical, physical, pharmaceutical and biological data about thousands of well-studied drugs and drug targets. It is primarily focused on providing the kind of detailed molecular data needed to facilitate drug and drug-target discovery. We believe DrugBank is one of the first examples of a dedicated, fully integrated drug/sequence/disease database. No doubt other databases of this kind will appear and it is expected that their development and dissemination will only improve the prospects for *in silico* drug-target discovery.

9 Analytical tools for drug-target discovery

For the most part we have focused almost exclusively on web-accessible databases aimed at facilitating drug-target discovery. Certainly the web is most often used as an information repository to upload and download facts and figures. However, the web can also be used to perform predictive or analytical operations. Using standard CGI (common gateway interface) forms, high-speed Internet connections, high-end servers and Java-based interface tools, it is now possible to deliver a number of advanced, interactive software services directly over the web. Rather than forcing users to buy, download and install programs (which is always rife with installation problems and platform compatibility issues), it is often easier for developers of specialized scientific software to have users access their software directly through the web.

Analytical web servers are now available to analyse or process a number of experimental processes commonly used in drug-target discovery. These include servers to process microarray or genechip data (Liu *et al.*, 2003; Brazma *et al.*, 2003; Knudsen *et al.*, 2003) and servers to process and annotate 2D gel data (Lemkin *et al.*, 1999; Young *et al.*, 2004). Interestingly, many of the analytical functions performed freely by these web servers are offered by commercial software companies at prices of $5000 to $10,000. Certainly commercial software generally has some important advantages (performance, support, user-friendliness) over freely available 'amateur' software. But it is important to note that many of the most significant innovations in proteomic or genomic analysis typically appear in freeware well before it makes its way into commercial packages. Likewise, with the rapid development and increased sophistication of web-based graphical user interfaces, many freely accessible web servers can closely match the performance and user-friendliness of some of the best commercial packages. Here we highlight two web servers that could be quite useful in both comparative genomics/proteomics and in drug-target discovery. One is for processing microarray data (GenePublisher) and the other is for processing 2D gel data (GelScape). Both are particularly simple to use and both are good examples of how the web can potentially be used to facilitate drug-target discovery by processing and analysing raw experimental data.

GenePublisher (Knudsen *et al.*, 2003) is a fully automated, web-based system for processing Affymetrix genechip experiments. It was originally developed by the Center for Biological Sequence Analysis in Denmark. While other web servers exist, such as NetAffx (Liu *et al.*, 2003) and ExpressionProfiler (Brazma *et al.*, 2003), what is particularly appealing about GenePublisher is its near total automation of the process and the high quality of its resulting reports and figures. In effect, with GenePublisher, one simply uploads their raw Affymetrix data, waits a few minutes and then a nearly finished paper (with methods, colour figures, tables and references – all in PDF format) is produced. The analytical procedures used in GenePublisher follow well-established peer-proven protocols and exploit a number of in-house servers, databases and tools uniquely available to the Danish Biological Sequence Analysis Center. Obviously, this kind of initial, automated analysis can be followed up with more detailed manual analyses or it can be used to suggest experiments for verification of the results. Nevertheless, in the field of drug-target discovery, these initial transcript analyses frequently point to important biomarkers, key pathways and critical triggers to diseases or disease processes.

GenePublisher conducts its GeneChip analysis using seven separate steps: (1) initial data/image analysis; (2) statistical and significance analysis; (3) K-nearest neighbour classification; (4) gene or transcript annotation (via GenBank, GO, TRANSPATH, KEGG, ProtFun); (5) K-means clustering; (6) promoter analysis; and (7) report generation. To begin the process, users must upload gzipped CEL files from an Affymetrix experiment or a 'genetable' of raw image analysis intensities. Once the data set is loaded, the initial data analysis including normalization, background correction, expression index calculation and visualization of chip-to-chip variation is performed using the Bioconductor affy package. By default, the qspline (quadratic spline) method is used for normalization. After this initial processing step, GenePublisher uses the R statistical programming environment to conduct a statistical and significance analysis. Principal component analysis (PCA) and hierarchical clustering is performed on the chip data to identify or display any obvious structure in the data. Additionally, t-tests (for 2-category cases) and ANOVA tests (for >2-category cases) are performed. A Bonferroni correction for multiple testing is then performed and the list of genes with significant differential expression is output, with calculated log fold changes. Once the differentially expressed genes have been identified the list is annotated with description of the genes and links to the LocusLink database, Gene Ontology (GO) labels, KEGG database links (if they exist), TRANSPATH (Krull *et al.*, 2003) links (if they exist) and putative functions predicted by homology to GO-annotated proteins or via ProtFun (Jensen *et al.*, 2002). ProtFun predicts protein function based on properties of the protein sequence as well as predicted features such as post-translational modification. In its fifth step of the analysis process, GenePublisher performs a hierarchical clustering on all the top-ranking genes. A K-means clustering is also run on the same highly ranked genes. The optimal number of clusters is chosen as the one which results in the smallest ratio of within-cluster to between-cluster variance. The program automatically chooses a colour scale to capture and display the spectrum of variation in the data. In the sixth step, GenePublisher performs a promoter analysis on the highest ranking genes by looking for known and unknown regulatory elements using three different methods: (1) a statistically based pattern searching tool, (2) a Gibbs sampler method and (3) direct comparisons to the TRANSFAC database (Matys *et al.*, 2003). Once the promoter analysis is complete GenPublisher summarizes what it did using a templated (fill-in-the-blank) report structure based on the analysis performed and parameters chosen. The report, complete with annotated tables and graphs, is converted to a PDF document and returned to the user via the web interface. Also returned is a table of normalized intensities and P-values of all genes in all experiments. A sample report can be downloaded from the GenePublisher website. Overall, this microarray analysis tool is really quite impressive and certainly simplifies many of the steps that are manually done using tedious and time-consuming steps required by most commercial software suppliers. It also allows researchers the time to think about and interpret their microarray data in biologically meaningful terms. This can go a long way to identifying and validating promising drug targets.

Just as with transcript-based gene chip methods, proteomics-based methods in drug-target discovery often depend on being able to compare the expression of proteins between two or more different conditions (diseased vs. healthy, treated vs. untreated). Because of its simplicity and its ability to separate and display even tiny amounts of protein, 2D gel electrophoresis has been and continues to be one of the most popular, low-cost methods to quantify and identify differential protein expression. While running 2D gels is a relatively simple experimental process, analysing 2D gels is not. In fact, gel analysis is normally conducted with fairly sophisticated image analysis and image manipulation software. Over the past decade a number of popular, commercial software packages have been introduced to facilitate digital gel analysis, including Melanie 4 (GeneBio), Phoretix 2D (Phoretix Inc.), ImageMaster 2D (Amersham), PDQuest (BioRad) and Gellab II (Scanalytics). These stand-alone programs support an impressive array of interactive image analysis and annotation routines. However, the high cost ($3000–$10,000) and restricted platform compatibility of many of these packages have led some groups to attempt to develop web-based freeware or groupware systems to handle certain key aspects of gel analysis, including, archiving (SWISS-PROT 2D – Appel *et al.*, 1999), comparison (Flicker – Lemkin, 1997) or interactive exploration (WebGel – Lemkin *et al.*, l999). Indeed, recent advances in Java applet, Java servlet and Java Server Page (JSP) technologies have proven that it is possible to perform very sophisticated, interactive image manipulation over the web. These advances actually motivated our development of a freely available web-based gel analysis system called GelScape.

GelScape (Young *et al.*, 2004) is designed to permit both 1D and 2D gel annotation (via MS analyses) as well as gel overlay and comparison. These kinds of functions are critical to being able to identify differential expression or differential modifications (phosphorylation, cleavage) of proteins. They are also critical for being able to identify potential drug targets, for both exogenous and endogenous diseases, from standard proteomic experiments. The GelScape server, itself, is composed of six functional windows : (1) the 'Load Gel' window; (2) the 'Grid&Axes' window; (3) the 'Annotate&View' window; (4) the 'Manipulate Gel' window; (5) the Gel 'Morph&Compare' window and (6) the 'GelBank' (gel archiving) window. Navigation to different windows is readily accessible by clicking on the various hyperlinked tabs located at the top of each window. Users begin a GelScape session by logging into the GelScape home page. Once logged in, the user is transferred to the Load Gel window. In this window, gel images (gif or jpg, 1D or 2D) may be uploaded (and later resized) from the user's local machine using simple HTML file browser. Hyperlinked thumbnail images of previously loaded or annotated gels are also displayed for easy reloading.

After loading the desired image, the user is transferred to the Annotate&View window. From this window, users may either interactively view their gel or navigate to other GelScape windows. If a gel is being annotated for the first time, users typically select the Grid&Axes window to define and/or interactively adjust the molecular weight and/or pH gradient scales. These scales are not only drawn on the gel (grid lines may be toggled off and on), but are also mapped onto the Annotate&View window which interactively displays these values on the image as the cursor is moved. Once the grid is set, the user normally navigates to the Annotate&View window. In this window, users may manually mark and annotate gel spots or bands by clicking on the spot or band of interest and filling in the appropriate text boxes. Alternatively, users may annotate spots or bands by entering the protein accession number (SwissProt or GenBank), or a peptide mass list (analysed by the PeptideSearch/WWW server). GelScape automatically uses web queries to retrieve the relevant sequence, protein name, pI and MW data, all of which are uploaded to the appropriate text boxes. Users may subsequently input additional comments, add associated mass spectrometry images or files, edit the uploaded text data or rename the spot.

In addition to manual peak picking, GelScape also supports automated peak picking and peak integration. Spots are determined and quantified by comparing their intensities against a user-adjustable threshold. If the user-defined threshold does not accurately detect the spots, the user may either undo or selectively modify the marked spots as desired. Selected spots or bands are marked with either a resizeable cross (2D gel) or a line (1D gel). Once marked, gel annotations can be viewed within the Annotate&View window by clicking on the labelled spots displayed on the image. Alternatively, a gel legend with abbreviated annotations for all spots is also accessible from this same window. File saving can be done on the web-server's allocated disc space or the HTML image map can be emailed to the user's preferred mail account.

In addition to these interactive viewing and annotation tools, GelScape also supports a variety of image editing and manipulation functions. For instance, in the Manipulate Gel window, users may convert between image formats, re-colour gel spots, display spots/bands as filled or empty regions, change the background colour, or convert the image to a transparent form for facile gel superposition. In the Morph&Compare window, users may perform multi-gel comparisons (including gel 'flickering' and direct gel overlay), image warping and gel-to-gel annotation transfer. After a gel image has been adjusted or warped, it may be overlaid with another gel image (transparent or opaque) or the annotations from a reference gel may be directly transferred to another gel. Both operations are automatically implemented using directional arrow buttons.

Overall, GelScape is an easy-to-use, web-based gel analysis system that permits facile, interactive annotation, comparison, manipulation and storage of protein gel images. It supports many of the features found in commercial, stand-alone gel analysis software including spot annotation, spot integration, gel warping, image resizing, HTML image mapping, image overlaying as well as the storage of gel image and gel annotation data standard formats. In this regard, GelScape could serve as an ideal, low-cost tool to facilitate drug target identification and discovery in budget-conscious proteomics laboratories. However, GelScape does have some limitations. Given the usual computational needs for image processing along with limits in Internet data transfer speeds, GelScape is not able to handle large (>1 Mbyte) image files. Likewise, certain restrictions on client-side accessibility limit what GelScape can offer in terms of analytical and interface structures. Nevertheless, GelScape does represent a good example of the kind of sophisticated analytical tools that can be built and that are now being offered over the web. As new Internet technologies evolve and as data transfer speeds accelerate, it is likely that almost anything we normally expect to do on a stand-alone computer will soon be possible through a web-based servlet or applet.

10 Conclusion

In this chapter we have described how drug-target discovery can be done *in silico* using a variety of web-accessible databases and software. Specifically we have explored five different classes of web resources: (1) sequence databases; (2) automated genome annotation tools; (3) text-mining tools; (4) integrated drug/sequence databases; and (5) web-based microarray and 2D gel analysis tools. Many of these web-based resources represent the products that arose, either directly or indirectly, from the Human Genome Project. They also represent the results of efforts by many scientists and national institutions to facilitate the sharing of biologically important data over the web. In addition to providing a brief survey and general assessment of many web-based tools, we have also shown how they can be used differently to assist in target discovery for either exogenous (infectious) diseases or endogenous (in-born or age-related) diseases. Indeed, we can now redraw *Figures 1* and *2*, which outlined *in silico* drug-target identification, and replace the generic processes originally illustrated in these figures with some very specific programs and databases (*Figures 7* and *8*).

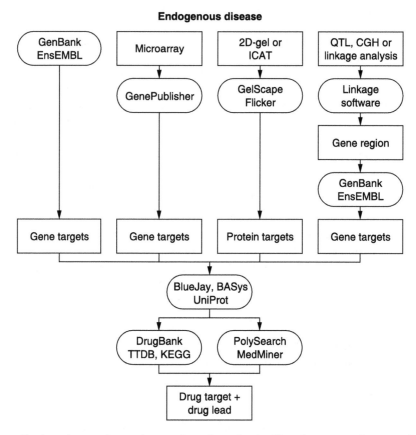

Figure 7. *Flow chart outlining the specific tools used in* in silico *drug-target discovery for endogenous (human) diseases. The rounded boxes represent software tools or databases while the square boxes represent experimental methods.*

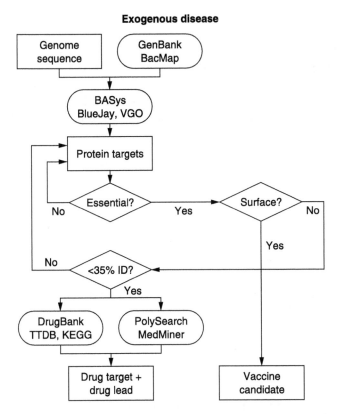

Figure 8. *Flow chart outlining the specific tools used in* in silico *drug-target discovery for exogenous (infectious) diseases. The rounded boxes represent software tools or databases while the square boxes represent experimental methods.*

Obviously the databases and analytical tools mentioned in *Figures 7* and *8* are not the only software or database choices. Many other options, both commercial and web-based, clearly exist. Likewise, many other options, both experimental and computational, also exist for drug-target discovery. These alternative approaches are presented and discussed in other chapters throughout this book. Overall, the particular emphasis placed on *in silico* target discovery in this chapter should not diminish the critical importance of solid, well-planned *in vivo* and *in vitro* experiments that are needed to stimulate and validate any *in silico* efforts. Indeed, *in silico* target discovery could not exist without being continuously complemented with, or motivated by, solid experimental research. Ultimately, what this chapter should illustrate is that the awareness and proper use of easily accessible web resources could make experimental efforts much easier, more focused and far more interpretable. In other words, an hour on the desk-top can often save a month at the bench-top.

Acknowledgement

The author wishes to acknowledge the support of Genome Prairie, a division of Genome Canada in developing some of the software and databases described in this chapter.

References

Altschul, S.F., Madden, T.L., Schaffer, A.A., Zhang, J., Zhang, Z., Miller, W. and Lipman, D.J. (1997) Gapped BLAST and PSI-BLAST: a new generation of protein database search programs. *Nucleic Acids Res* **25**: 3389–3402.

Andrade, M.A., Brown, N.P., Leroy, C., Hoersch, S., de Daruvar, A., Reich, C., Franchini, A., Tamames, J., Valencia, A., Ouzounis, C. and Sander, C. (1999) Automated genome sequence analysis and annotation. *Bioinformatics* **15**: 391–412.

Appel, R.D., Bairoch, A. and Hochstrasser, D.F. (1999) 2-D databases on the World Wide Web. *Methods Mol Biol* **112**: 383–391.

Baasiri, R.A., Glasser, S.R., Steffen, D.L. and Wheeler, D.A. (1999) The breast cancer gene database: a collaborative information resource. *Oncogene* **18**: 7958–7965.

Bairoch, A., Apweiler, R., Wu, C.H., Barker, W.C., Boeckmann, B., Ferro, S., Gasteiger, E., Huang, H., Lopez, R., Magrane, M., Martin, M.J., Natale, D.A., O'Donovan, C., Redaschi, N. and Yeh, L.S. (2005) The Universal Protein Resource (UniProt). *Nucleic Acids Res* **33(Database issue):** D154–159.

Bentley, D.R. (2000) The Human Genome Project – an overview. *Med Res Rev* **20**: 189–196.

Brazma, A., Parkinson, H., Sarkans, U., Shojatalab, M., Vilo, J., Abeygunawardena, N., Holloway, E., Kapushesky, M., Kemmeren, P., Lara, G.G. *et al.* (2003) ArrayExpress — a public repository for microarray gene expression data at the EBI. *Nucleic Acids Res* **31**: 68–71.

Brooksbank, C., Cameron, G. and Thornton, J. (2005) The European Bioinformatics Institute's data resources: towards systems biology. *Nucleic Acids Res* **33 (Database issue):** D46–53.

Butte, A. (2002) The use and analysis of microarray data. *Nat Rev Drug Discov* **1**: 951–960.

Buysse, J.M. (2001) The role of genomics in antibacterial target discovery. *Curr Med Chem* **8**: 1713–1726.

Carlton, J. (2003) The *Plasmodium vivax* genome sequencing project. *Trends Parasitol* **19**: 227–231.

Chan, P.F., Macarron, R., Payne, D.J., Zalacain, M. and Holmes, D.J. (2002) Novel antibacterials: a genomics approach to drug discovery. *Curr Drug Targets Infect Disord* **2**: 291–308.

Chaussabel, D. (2004) Biomedical literature mining: challenges and solutions in the 'omics' era. *Am J Pharmacogenomics* **4**: 383–393.

Chen, X., Ji, Z.L. and Chen, Y.Z. (2002) TTD: Therapeutic Target Database. *Nucleic Acids Res* **30**: 412–415.

Clarke, P.A., te Poele, R. and Workman, P. (2004) Gene expression microarray technologies in the development of new therapeutic agents. *Eur J Cancer* **40**: 2560–2591.

Cohen, A.M. and Hersh, W.R. (2005) A survey of current work in biomedical text mining. *Brief Bioinform* **6**: 57–71.

Comess, K.M. and Schurdak, M.E. (2004) Affinity-based screening techniques for enhancing lead discovery. *Curr Opin Drug Discov Devel* **7**: 411–416.

Darvasi, A. (2005) Dissecting complex traits: the geneticists' 'Around the world in 80 days'. *Trends Genet* **21**: 373–376.

Delcher, A.L., Harmon, D., Kasif, S., White, O., Salzberg, S.L. (1999) Imporved microbial gene identification with GLIMMER. *Nucleic Acids Res* Dec. 1; **27(23)**: 4636–4641.

Deshpande, N., Addess, K.J., Bluhm, W.F., Merino-Ott, J.C., Townsend-Merino, W., Zhang, Q., Knezevich, C., Xie, L., Chen, L., Feng, Z., Green, R.K., Flippen-Anderson J.L., Westbrook, J., Berman, H.M. and Bourne, P.E. (2005) The RCSB Protein Data Bank: a redesigned query system and relational database based on the mmCIF schema. *Nucleic Acids Res* **33(Database issue)**: D233–237.

Fischer, H.P. (2005) Towards quantitative biology: integration of biological information to elucidate disease pathways and to guide drug discovery. *Biotechnol Annu Rev* **11**: 1–68.

Fishman, R.H. (1996) Bioinformatics speeds HIV-1 drug development. *Lancet* **348**: 1648.

Fraser, C.M. and Rappuoli, R. (2005) Application of microbial genomic science to advanced therapeutics. *Annu Rev Med* **56**: 459–474.

Frazer, K.A., Elnitski, L., Church, D.M., Dubchak, I. and Hardison, R.C. (2003) Cross-species sequence comparisons: a review of methods and available resources. *Genome Res* **13**: 1–12.

Gaasterland, T. and Sensen, C.W. (1996) MAGPIE: automated genome interpretation. *Trends Genet* **12**: 76–88.

Griffin, J.L. (2003) Metabonomics: NMR spectroscopy and pattern recognition analysis of body fluids and tissues for characterisation of xenobiotic toxicity and disease diagnosis. *Curr Opin Chem Biol* **7**: 648–654.

Hammond, M.P. and Birney, E. (2004) Genome information resources – developments at Ensembl. *Trends Genet* **20**: 268–272.

Hamosh, A., Scott, A.F., Amberger, J.S., Bocchini, C.A. and McKusick, V.A. (2005) Online Mendelian Inheritance in Man (OMIM), a knowledgebase of human genes and genetic disorders. *Nucleic Acids Res* **33(Database issue)**: D514–517.

Harris, N.L. (1997) Genotator: a workbench for sequence annotation. *Genome Res* **7**: 754–762.

Harris, S. and Foord, S.M. (2000) Transgenic gene knock-outs: functional genomics and therapeutic target selection. *Pharmacogenomics* **1**: 433–443.

Hatfield, C.L., May, S.K. and Markoff, J.S. (1999) Quality of consumer drug information provided by four Web sites. *Am J Health Syst Pharm* **56**: 2308–2311.

Hewett, M., Oliver, D.E., Rubin, D.L., Easton, K.L., Stuart, J.M., Altman, R.B. and Klein, T.E. (2002) PharmGKB: the Pharmacogenetics Knowledge Base. *Nucleic Acids Res* **30**: 163–165.

Hiscock, D. and Upton, C. (2000) Viral Genome DataBase: storing and analyzing genes and proteins from complete viral genomes. *Bioinformatics* **16**: 484–485.

Hoffmann, R. and Valencia, A. (2005) Implementing the iHOP concept for navigation of biomedical literature. *Bioinformatics* **21(Suppl 2)**: ii252–ii258.

Hopkins, A.L and Groom, C.R. (2002) The druggable genome *Nat Rev Drug Discov* **1**: 727–730.

Inazawa, J., Inoue, J. and Imoto, I. (2004) Comparative genomic hybridization (CGH)-arrays pave the way for identification of novel cancer-related genes. *Cancer Sci* 95: 559–563.

Jeffery, D.A. and Bogyo, M. (2003) Chemical proteomics and its application to drug discovery. *Curr Opin Biotechnol* 14: 87–95.

Jensen,L. J., Gupta, R., Blom, N., Devos, D., Tamames, J., Kesmir, C., Nielsen, H., Staerfeldt, H.H., Rapacki, K., Workman, C. *et al.* (2002) *Ab initio* reduction of human orphan protein function from post-translational modifications and localization features. *J Mol Biol* 319: 1257–1265.

Jonsdottir, S.O., Jorgensen, F.S. and Brunak, S. (2005) Prediction methods and databases within chemoinformatics: emphasis on drugs and drug candidates. *Bioinformatics* 21: 2145–2160.

Kanehisa, M., Goto, S., Kawashima, S., Okuno, Y. and Hattori, M. (2004) The KEGG resource for deciphering the genome. *Nucleic Acids Res* 32(Database issue): D277–280.

Karolchik, D., Baertsch, R., Diekhans, M., Furey, T.S., Hinrichs, A., Lu, Y.T., Roskin, K.M., Schwartz, M., Sugnet, C.W., Thomas, D.J., Weber, R.J., Haussler, D., Kent, W.J. (2003) The UCSC Genome Browser Database. *Nucleic Acids Res* 31: 51–54.

Knudsen, S., Workman, C., Sicheritz-Ponten, T. and Friis, C. (2003) GenePublisher: Automated analysis of DNA microarray data. *Nucleic Acids Res* 31: 3471–3476.

Kramer, R. and Cohen, D. (2004) Functional genomics to new drug targets. *Nat Rev Drug Discov* 3: 965–972.

Krull, M., Voss, N., Choi, C., Pistor, S., Potapov, A. and Wingender, E. (2003) TRANSPATH: an integrated database on signal transduction and a tool for array analysis. *Nucleic Acids Res,* 31: 97–100.

LaBaer, J. (2003) Mining the literature and large datasets. *Nat Biotechnol* 21: 976–977.

Lahm, A., Yagnik, A., Tramontano, A. and Koch, U. (2002) Hepatitis C virus proteins as targets for drug development: the role of bioinformatics and modelling. *Curr Drug Targets* 3: 281–296.

Lemkin, P.F. (1997) Comparing two-dimensional electrophoretic gel images across the Internet. *Electrophoresis* 18: 461–470.

Lemkin, P.F., Myrick, J.M., Lakshmanan, Y., Shue, M.J., Patrick, J.L., Hornbeck, P.V., Thornwal, G.C. and Partin, A.W. (1999) Exploratory data analysis groupware for qualitative and quantitative electrophoretic gel analysis over the Internet-WebGel. *Electrophoresis* 18: 3492–3507.

Lindsay, M.A. (2003). Target discovery. *Nat Rev Drug Discov* 2: 831–838.

Liu, G., Loraine, A.E., Shigeta, R., Cline, M., Cheng, J., Valmeekam, V., Sun, S., Kulp, D. and Siani-Rose, M.A. (2003) NetAffx: Affymetrix probesets and annotations. *Nucleic Acids Res* 31: 82–86.

Maglott, D., Ostell, J., Pruitt, K.D. and Tatusova, T. (2005) Entrez Gene: gene-centered information at NCBI. *Nucleic Acids Res* 33(Database issue): D54–58.

Manber, U. and Bigot, P. (1997) *USENIX Symposium on Internet Technologies and Systems (NSITS'97)*, Monterey, California, pp. 231–239.

Matys, V., Fricke, E., Geffers, R., Gossling, E., Haubrock, M., Hehl, R., Hornischer, K., Karas, D., Kel, A.E., Kel-Margoulis, O.V. *et al.* (2003) TRANSFAC: transcriptional regulation, from patterns to profiles. *Nucleic Acids Res* 31: 374–378.

Meyer, F., Goesmann, A., McHardy, A.C., Bartels, D., Bekel, T., Clausen, J., Kalinowski, J., Linke, B., Rupp, O., Giegerich, R. and Puhler, A. (2003) GenDB – an open source genome annotation system for prokaryote genomes. *Nucleic Acids Res* **31**: 2187–2195.

O'Donovan, C., Martin, M.J., Gattiker, A., Gasteiger, E., Bairoch, A. and Apweiler, R. (2002) High-quality protein knowledge resource: SWISS-PROT and TrEMBL. *Brief Bioinform* **3**: 275–284.

Onyango, P. (2004) The role of emerging genomics and proteomics technologies in cancer drug target discovery. *Curr Cancer Drug Targets* **4**: 111–124.

Orth, A.P., Batalov, S., Perrone, M. and Chanda, S.K. (2004) The promise of genomics to identify novel therapeutic targets. *Expert Opin Ther Targets* **8**: 587–596.

Paine, K. and Flower, D.R. (2002) Bacterial bioinformatics: pathogenesis and the genome. *J Mol Microbiol Biotechnol* **4**: 357–365.

Patton, W.F. (2002) Detection technologies in proteome analysis. *J Chromatogr B Analyt Technol Biomed Life Sci* **771**: 3–31.

Payne, D.J., Gwynn, M.N., Holmes, D.J. and Rosenberg, M. (2004) Genomic approaches to antibacterial discovery. *Methods Mol Biol* **266**: 231–259.

Peterson, J.D., Umayam, L.A., Dickinson, T., Hickey, E.K. and White, O. (2001) The Comprehensive Microbial Resource. *Nucleic Acids Res* **29**: 123–125.

Rebhan, M., Chalifa-Caspi, V., Prilusky, J. and Lancet, D.(1998) GeneCards: a novel functional genomics compendium with automated data mining and query reformulation support. *Bioinformatics* **14**: 656–664.

Reiter, L.T. and Bier, E. (2002) Using *Drosophila melanogaster* to uncover human disease gene function and potential drug target proteins. *Expert Opin Ther Targets* **6**: 387–399.

Riley, M.L., Schmidt, T., Wagner, C., Mewes, H.W. and Frishman, D. (2005) The PEDANT genome database in 2005. *Nucleic Acids Res* **33(Database issue)**: D308–310.

Ryan, T.E. and Patterson, S.D. (2002) Proteomics: drug target discovery on an industrial scale. *Trends Biotechnol* **20(12 Suppl)**: S45–51.

Southan, C. (2004) Has the yo-yo stopped? An assessment of human protein-coding gene number. *Proteomics* **4**: 1712–1726.

Stenson, P.D., Ball, E.V., Mort, M., Phillips, A.D., Shiel, J.A., Thomas, N.S., Abeysinghe, S., Krawczak, M. and Cooper, D.N. (2003) Human Gene Mutation Database (HGMD): 2003 update. *Hum Mutat* **21**: 577–581.

Stothard, P., Van Domselaar, G., Shrivastava, S., Guo, A., O'Neill, B., Cruz, J., Ellison, M. and Wishart, D.S. (2005) BacMap: an interactive picture atlas of annotated bacterial genomes. *Nucleic Acids Res* **33(Database issue)**: D317–320.

Sundararaj, S., Guo, A., Habibi-Nazhad, B., Rouani, M., Stothard, P., Ellison, M. and Wishart, D.S. (2004) The CyberCell Database (CCDB): a comprehensive, self-updating, relational database to coordinate and facilitate in silico modeling of *Escherichia coli*. *Nucleic Acids Res* **32(Database issue)**: D293–295.

Szafron, D, Lu, P, Greiner, R, Wishart, DS, Poulin, B, Eisner, R, Lu, Z, Anvik, J, Macdonell, C, Fyshe, A, Meeuwis, D. (2004) Proteome Analyst: custom predictions with explanations in a web-based tool for high-throughput proteome annotations. *Nucleic Acids Res* **32(Web Server issue)**: W365–371.

Tanabe, L., Scherf, U., Smith, L.H., Lee, J.K., Hunter, L. and Weinstein, J.N. (1999) MedMiner: an Internet text-mining tool for biomedical information, with application to gene expression profiling. *Biotechniques* **27**: 1210–1217.

Turinsky, A.L., Ah-Seng, A.C., Gordon, P.M., Stromer, J.N., Taschuk, M.L., Xu, E.W. and Sensen, C.W. (2005) Bioinformatics visualization and integration with open standards: the Bluejay genomic browser. *In Silico Biol* **5**: 187–198.

Van Domselaar, G.H., Stothard, P., Shrivastava, S., Cruz, J.A., Guo, A., Dong, X., Lu, P., Szafron, D., Greiner, R. and Wishart, D.S. (2005) BASys: a web server for automated bacterial genome annotation. *Nucleic Acids Res* **33**(Web Server issue): W455–459.

Voorhoeve, P.M. and Agami, R. (2003) Knockdown stands up. *Trends Biotechnol* **21**: 2–4.

Walgren, J.L. and Thompson, D.C. (2004) Application of proteomic technologies in the drug development process. *Toxicol Lett* **149**: 377–385.

Weininger, D. (1988) SMILES 1. Introduction and Encoding Rules. *J Chem Inf Comput Sci* **28**: 31–38.

Wheeler, D.L., Barrett, T., Benson, D.A., Bryant, S.H., Canese, K., Church, D.M., DiCuccio, M., Edgar, R., Federhen, S., Helmberg, W., Kenton, D.L., Khovayko, O., Lipman, D.J., Madden, T.L., Maglott, D.R., Ostell, J., Pontius, J.U., Pruitt, K.D., Schuler, G.D., Schriml, L.M., Sequeira, E., Sherry, S.T., Sirotkin, K., Starchenko, G., Suzek, T.O., Tatusov, R., Tatusova, T.A., Wagner, L. and Yaschenko, E. (2005) Database resources of the National Center for Biotechnology Information. *Nucleic Acids Res* **33**(Database issue): D39–45.

Wishart, D.S. (2005) Bioinformatics in drug development and assessment. *Drug Metab Rev* **37**: 279–310.

Young, N., Chang, Z. and Wishart, D.S. (2004) GelScape: a web-based server for interactively annotating, manipulating, comparing and archiving 1D and 2D gel images. *Bioinformatics* **20**: 976–978.

Zon, L.I. and Peterson, R.T. (2005) In vivo drug discovery in the zebrafish. *Nat Rev Drug Discov* **4**: 35–44.

Index

Note: page numbers in *italics* refer to figures, those in **bold** refer to tables.

Milton Keynes UK
Ingram Content Group UK Ltd.
UKHW040055071024
449327UK00019B/584

9 780367 389734